Prepare for Stat

with ArgoPrep

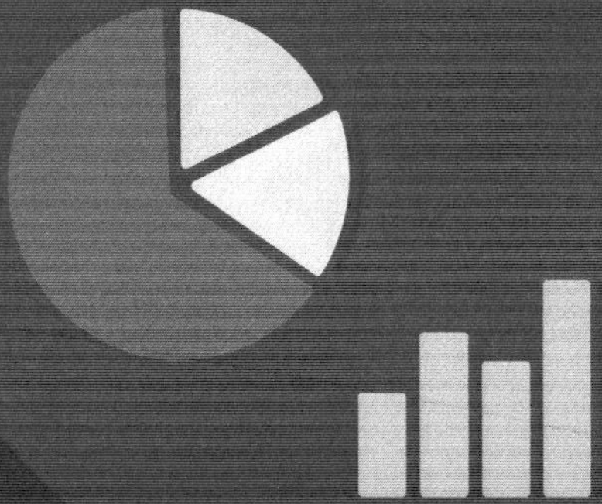

5th Grade Math Practice Book

8 Full-Length Simulated Common Core Math Tests

ArgoPrep is one of the leading providers of supplemental educational products and services. We offer affordable and effective test prep solutions to educators, parents and students. Learning should be fun and easy! To access more resources visit us at www.argoprep.com.

Our goal is to make your life easier, so let us know how we can help you by e-mailing us at: info@argoprep.com.

ISBN: 9781951048440
Published by ArgoPrep.

- ArgoPrep is a recipient of the prestigious **Mom's Choice Award**.
- ArgoPrep also received the 2019 **Seal of Approval** from Homeschool.com for our award-winning workbooks.
- ArgoPrep was awarded the 2019 **National Parenting Products Award, Gold Medal Parent's Choice Award** and **the Tillywig Brain Child Award.**

BOOKS BY ARGOPREP

KIDS SUMMER ACADEMY SERIES

ArgoPrep's Kids Summer Academy series helps prevent summer learning loss and gets students ready for their new school year by reinforcing core foundations in math, english and science. Our workbooks also introduce new concepts so students can get a head start and be on top of their game for the new school year!

SUMMER ACTIVITY PLAYGROUND SERIES

Summer Activity Playground is another summer series that is designed to prevent summer learning loss and prepares students for the new school year. Students will be able to practice math, ELA, science, social studies and more! This is a new released series that offers the latest aligned learning standards for each grade.

Table of contents

How to use the book

This workbook contains eight full-length practice exams that simulate state exams for **Grade 5**. The goal of this workbook is to provide mock state tests so students can increase confidence and test scores during actual test day. The exams mirror New York State Exams and reflect the latest learning standards tested for the **2020-2021** school year. If you are a teacher using this workbook, you can find the common core standards for each question in the answer key.

ArgoPrep recommends users of this workbook to simulate real testing conditions to provide the maximum benefit. Students should aim to complete an entire session in a sitting just like in the state exam. Please provide students with extra scrap paper while working through the exams.

For a detailed overview of the Common Core State Standards for 5th grade, please visit: www.corestandards.org/Math/Content/5/introduction/

Have any questions? Contact us at: **info@argoprep.com**

CONVERSIONS

1 mile = 5,280 feet
1 mile = 1,760 yards

1 pound = 16 ounces
1 ton = 2,000 pounds

1 cup = 8 fluid ounces
1 pint = 2 cups
1 quart = 2 pints
1 gallon = 4 quarts
1 liter = 1,000 cubic centimeters

FORMULAS

Right Rectangular Prism V = Bh or V = lwh

Test 1: Session 1

Tips: Be sure to read each question carefully and work through the problem step by step before choosing the correct answer.

Suggested Time:

Session 1: 55 to 65 Minutes
Session 2: 60 to 75 minutes

Students are **NOT** permitted to use a calculator.

1. Maria is baking cakes. She is making five batches. Each batch calls for $\frac{1}{3}$ cup of baking soda. How much baking soda is she using for all of the cakes?

A. $5\frac{1}{3}$ cups **B.** $\frac{4}{3}$ cups **C.** $1\frac{1}{3}$ cups **D.** $1\frac{2}{3}$ cups

2. Which number is represented by the following expression?

(4 x 1,000) + (2 x 100) + (1 x 10) + (5 x 1) + (3 x .1) + (8 x .01)

A. 4,215.38
B. 4,125.38
C. 4,125.83
D. 4,215.83

3. Rex spent $\frac{1}{2}$ an hour each day practicing piano for 5 days. Thomas spent $\frac{2}{3}$ an hour each day practicing guitar for 4 days. Who spent more time practicing? How much more time did they spend?

A. Thomas spent $\frac{1}{6}$ of an hour more practicing
B. Thomas spent $1\frac{1}{2}$ hours more practicing
C. Rex spent $2\frac{2}{3}$ hours more practicing
D. Rex spent 1 hour more practicing

4. Three friends are enjoying leftover cupcakes. There are only 2 cupcakes left, but they want to each have an equal amount. How much can they each have?

A. $\frac{2}{3}$ of a cupcake

B. $\frac{1}{2}$ of a cupcake

C. $\frac{3}{4}$ of a cupcake

D. $\frac{1}{5}$ of a cupcake

5. Cecilia is setting up a book shop. She has 12 bookshelves and 4,140 books. She is trying to determine how many books fit on each bookshelf using an area model. What is her first partial quotient?

A. 200

B. 300

C. 400

D. 600

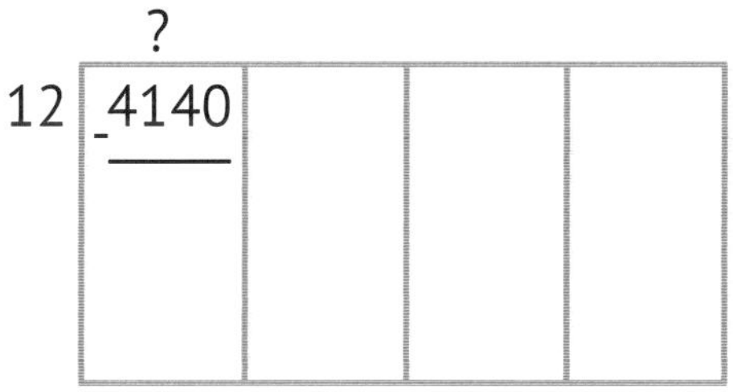

6. Amal has $\frac{3}{4}$ lbs of flour. He needs to divide it equally between two different recipes. How much flour will he use for each recipe?

A. $\frac{8}{3}$ lbs of flour

B. 6 cups of flour

C. $\frac{3}{8}$ lbs of flour

D. 8 cups of flour

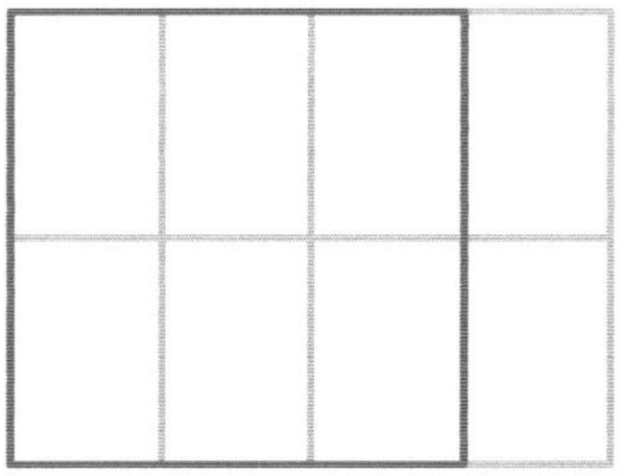

GO ON

7. Sarah compares the length of different vines.

Vine A	$7\frac{1}{2}$ inches
Vine B	$9\frac{3}{10}$ inches
Vine C	$7\frac{3}{4}$ inches

How much longer is Vine C than Vine A?

A. 1 inch

B. 2 inches

C. $2\frac{1}{2}$ inches

D. $\frac{1}{4}$ inch

8. Last year a company sold 5,125 T-shirts in their online store. Each of the T-shirts cost $7. How much money did the company make selling T-shirts?

A. $3,587

B. $35,885

C. $3,975

D. $35,875

9. Levi is playing with toy trains. He wants to connect all of the trains together to create one giant train. Using the line plot below, determine the length of all trains when they are put together.

A. $3\frac{1}{4}$ ft

B. $5\frac{1}{2}$ ft

C. $6\frac{1}{4}$ ft

D. $7\frac{1}{2}$ ft

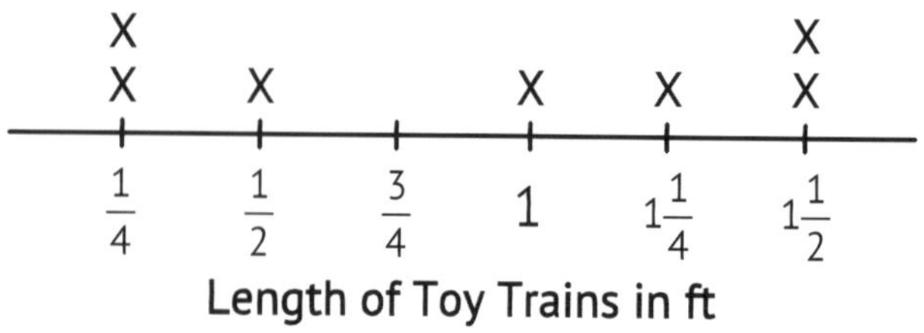

10. Look at the line plot from previous question. How much longer is the longest train compared to the shortest train?

A. $1\frac{1}{4}$ ft

B. 1 ft

C. $1\frac{1}{2}$ ft

D. 2 ft

11. Kinsley is cutting strips of fabric for a costume. She has 3 yards of fabric. She wants to cut strips that are 3 inches wide. How many 3 inch strips of fabric can she cut?

A. 108 strips
B. 48 strips
C. 36 strips
D. 12 strips

12. Matthew is measuring the weight of rock samples. The last one he measured weighed 5.284 ounces. What is the weight rounded to the tenths place?

A. 5.2 ounces
B. 5.3 ounces
C. 5.28 ounces
D. 5.29 ounces

GO ON

13. Greg is feeding horses on his farm. He has 6 bales of hay. Each horse eats $\frac{1}{3}$ of a bale. How many horses can Greg feed?

A. 2 horses
B. 3 horses
C. 12 horses
D. 18 horses

14. Students from all over the northeast traveled to a Lego convention in New York. Each bus was filled with 95 students. There were 315 buses in all. How many students attended the Lego convention?

A. 29,935
B. 29,915
C. 29,925
D. 30,025

15. Which coordinates describe the position of the point on the coordinate plane?

A. 6,8
B. 8,6
C. (6,8)
D. (8,6)

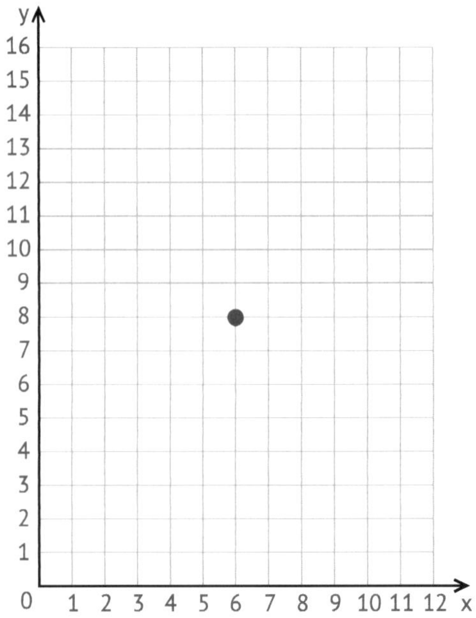

16. What is the cubic volume of the figure?

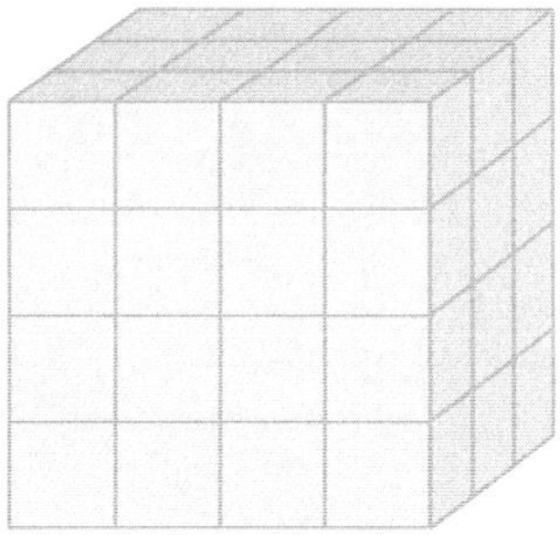

A. 48 cubic units
B. 16 cubic units
C. 44 cubic units
D. 11 cubic units

17. A group of friends were comparing their vacation and travel. They organized their data on the coordinate plane. How many more cities has Chuck visited than Danielle?

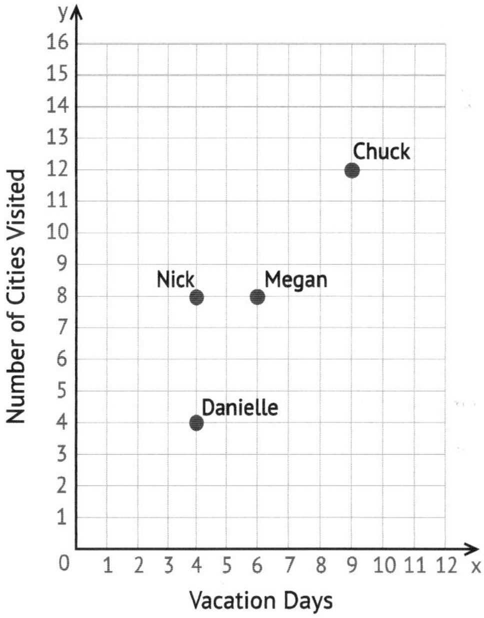

A. 8 cities
B. 12 cities
C. 4 cities
D. 6 cities

18. Which of the following expressions is equal to 713?

A. $713 \times \frac{1}{2}$ **B.** $713 \times \frac{5}{5}$ **C.** $713 \times \frac{10}{6}$ **D.** $713 \times \frac{1}{713}$

GO ON

19. There are 453 people playing in a national soccer league. Each of the four regions has an equal number of players. Which multiplication can help estimate the number of players in each region?

A. 4 x 10 = 40

B. 4 x 20 = 80

C. 4 x 100 = 400

D. 4 x 200 = 800

20. Isaac is building a sandcastle. He is using two stacked boxes and filling them with sand. How many cubic centimeters of sand will be in his sandcastle?

A. 34 cm^3

B. 354 cm^3

C. 364 cm^3

D. 254 cm^3

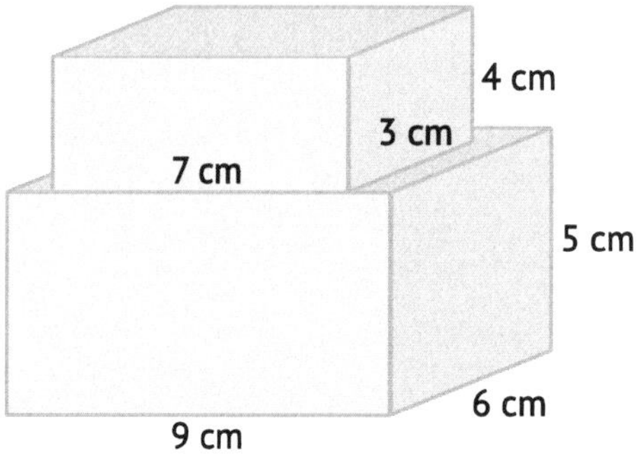

21. Last year Ramsey earned $50 mowing lawns during the summer. This year he was able to earn $5,000 mowing lawns! How much more money did he earn this summer?

A. $\frac{1}{10}$ as much money

B. $\frac{1}{100}$ as much money

C. 10 times as much money

D. 100 times as much money

22. Stephanie is comparing the size of planets. Which of the following expressions would yield the greatest mass?

A. 3.28×10^{23}

B. 4.86×10^{24}

C. 3.28×10^{24}

D. 4.86×10^{23}

Alan was swimming laps in the pool and used a stopwatch to record his times. He swam his first lap in 2.514 minutes. He swam his second lap in 2.13 minutes. He swam his third lap in 3.507 minutes. After swimming, he went on a run for 15.832 minutes.

23. How much time did he spend swimming laps?

A. 8.151 minutes

B. 8.124 minutes

C. 8.214 minutes

D. 9.051 minutes

24. How much more time did he spend running than swimming?

A. 6.781 minutes

B. 7.691 minutes

C. 7.701 minutes

D. 7.681 minutes

GO ON

25. Stephen and Juan have leftover cake from a party. Stephen has $\frac{2}{8}$ of his cake left. Juan has $\frac{2}{3}$ of his cake left. How much leftover cake do they have altogether?

A. $\frac{4}{24}$

B. $\frac{22}{24}$

C. $\frac{4}{11}$

D. $\frac{6}{16}$

26. Leo is painting walls in his house. He has two gallons of paint. He uses $\frac{1}{4}$ of a gallon to paint the first wall. He uses $\frac{4}{5}$ of a gallon to paint the second wall. About how much paint did he use?

A. Less than $\frac{1}{2}$ gallon

B. Exactly $\frac{1}{2}$ gallon

C. Approximately 1 gallon

D. Approximately 2 gallons

27. Molly had $\frac{2}{3}$ of a pizza leftover. She gave $\frac{1}{2}$ of it to her friend. How much pizza did her friend get?

A. 1 piece

B. $\frac{3}{5}$ of a the pizza

C. $\frac{1}{2}$ of the pizza

D. $\frac{2}{6}$ of the pizza

28. Evaluate the expression:

7 x (15 - 5) + 2 = ______

A. 70
B. 107
C. 72
D. 105

29. What can a cubic unit be used to measure?

A. Weight
B. Length and Width
C. Time
D. Length, Width, and Height

30. Sandra is mixing concrete. She needs 5 gallons of water, but she only has a 1 cup measuring cup. How many cups of water are in 5 gallons?

A. 80 cups
B. 60 cups
C. 4 cups
D. 25 cup

GO ON

Session 2

31. Tommy is building a block tower. He has 8 blocks. Each block is $\frac{3}{4}$ inch tall. How tall would his tower be?

A. 3 inches

B. $4\frac{1}{2}$ inches

C. 6 inches

D. 8 inches

32. What is the value of 7.509 x 100?

A. .7509

B. 75.09

C. 750.9

D. 7,590

33. Round 41.873 to the nearest tenth.

A. 41.9

B. 40

C. 41.87

D. 42

GO ON

34. Solve.

$$\begin{array}{r} 5{,}129 \\ \times \quad 7 \\ \hline \end{array}$$

A. 35,903

B. 35,893

C. 36,903

D. 35,904

35. Varun is separating picture frames into boxes. He has 418 picture frames to pack. He has 8 boxes and he wants to put an even number of frames in each box. Which equation helps him estimate the number of frames he can put in each box?

A. $\frac{40{,}000}{8} = 5{,}000$

B. $\frac{4{,}000}{8} = 500$

C. $\frac{400}{8} = 50$

D. $\frac{40}{8} = 5$

36. Which makes the expression true?

.437 < ______

A. Four hundred two thousandths

B. Four hundred forty thousandths

C. Four hundred one thousandths

D. Four hundred thirty-one thousandths

37. An art teacher bought 3 lbs of clay for her class. Each student needs 2 oz of clay for their project. How many students will have enough clay?

A. 24 students

B. 48 students

C. 36 students

D. 5 students

38. Which expression matches the phrase below?

Four times the sum of eighteen and three

A. 4 x 18 + 3

B. 4 x 18 - 3

C. 4 x (18 + 3)

D. (4 x 18) + 3

39. Harold is making treat bags for a birthday party. He has 942 pieces of candy. He has 12 bags. He wants to make sure each bag has an equal amount of candy. How many pieces will be in each bag? Will there be any left over?

Show your work.

____

____

GO ON

40. Use the number line to divide.

$\frac{.8}{2}$ = ______

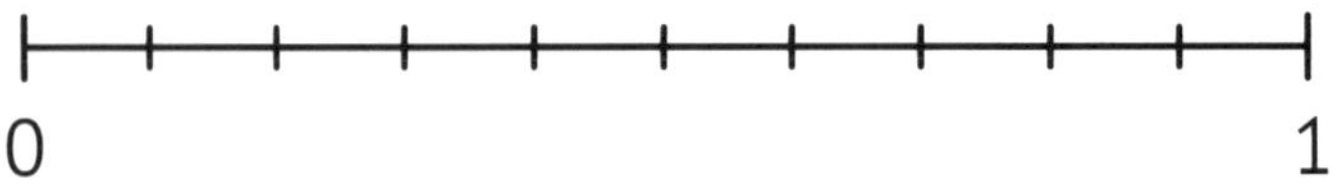

41. George is adding oil to his cars. His first car needs $\frac{3}{8}$ quart of oil. His second car needs $\frac{3}{4}$ quart of oil. He thinks he needs to buy one quart of oil. Is he correct or incorrect?

Explain your thinking.

____

42. Shade the figure below to show $\frac{1}{4} \div 6$

$\frac{1}{4} \div 6$ = ______

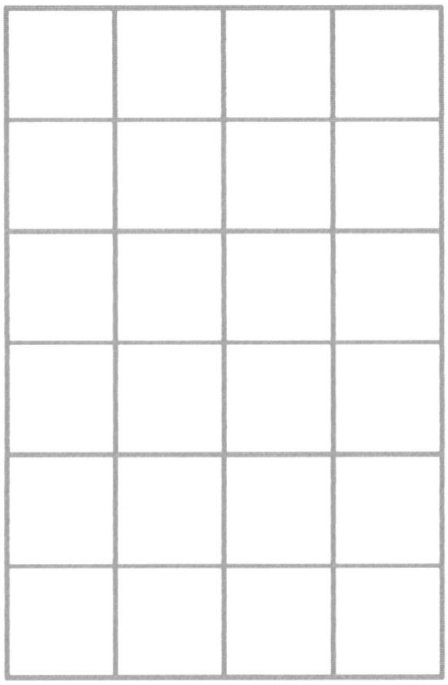

Marco is studying types of beetles. He measured their lengths and recorded them on the table below.

Ladybird beetle	$\frac{1}{4}$ cm, $\frac{1}{2}$ cm, $\frac{1}{4}$ cm	Brown beetle	$1\frac{3}{4}$ cm, 2 cm, $1\frac{1}{2}$ cm
Stag beetle	$\frac{3}{4}$ cm, $1\frac{1}{4}$ cm	Bark beetle	$\frac{2}{8}$ cm, $1\frac{1}{4}$ cm

43. Use the line below to create a line plot for the data.

44. How much longer is the largest beetle compared to the smallest beetle?

A. $1\frac{3}{4}$ cm **B.** $\frac{3}{4}$ cm **C.** $\frac{1}{2}$ cm **D.** 1 cm

45. Use the table below to determine the rule. Fill in the missing data points to create coordinate points. Graph the coordinate points on the plane

x	**y**	**(x, y)**
1	3	(1,3)
2	6	(2,6)
	9	(__, 9)
4		(4, __)
	15	(__,15)
Rule = ______		

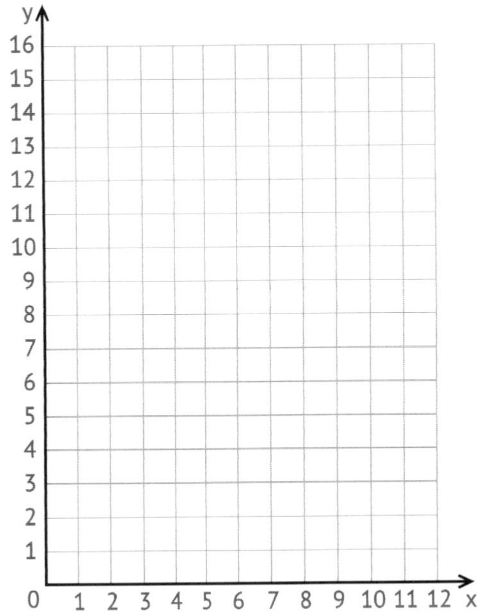

GO ON

Test 2: Session 1

Tips:

Be sure to read each question carefully and work through the problem step by step before choosing the correct answer.

Suggested Time:

Session 1: 55 to 65 Minutes
Session 2: 60 to 75 minutes

Students are **NOT** permitted to use a calculator.

1. Alex is separating Halloween candy into 4 equal piles with his friends. Altogether they collected 321 pieces of candy. Use the area model below to determine how much candy each person will get.

A. 150

B. 150 R 1

C. 80

D. 80 R 1

2. A dog shelter is collecting bags of dog treats. They have 452 bags. Each bag has 25 treats. How many treats have they collected so far?

A. 1,400 treats

B. 11,400 treats

C. 11,300 treats

D. 1,300 treats

3. Mike is eating pie at Thanksgiving. He eats $\frac{2}{8}$ of a pumpkin pie, then he eats $\frac{5}{6}$ of a pecan pie. He is stuffed! Which is closest to the amount of pie he ate?

A. Less than $\frac{1}{2}$ pie

B. About $\frac{1}{2}$ a pie

C. More than $\frac{1}{2}$ pie

D. About 1 whole pie

GO ON

4. Jordan built a new fence for his dog.

What **best** describes the attributes of the shape?

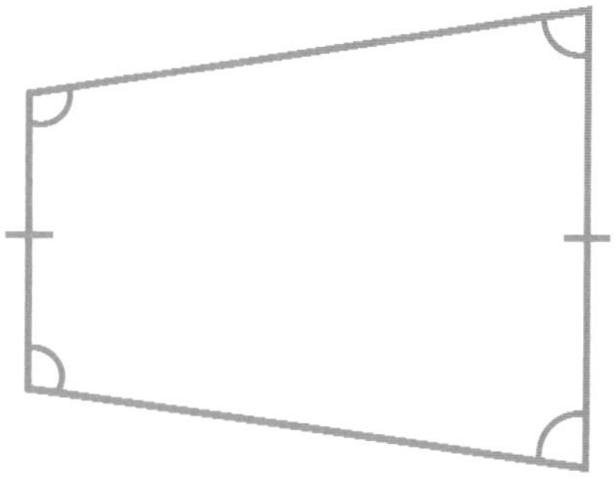

A. It is a rectangle.
B. It is a parallelogram.
C. It is a quadrilateral
D. It is a quadrilateral and a parallelogram.

5. Isaac is connecting strands of Christmas lights. The first strand is $3\frac{2}{3}$ feet long. The second strand is $4\frac{1}{2}$ feet long. How long will the combined strand be?

A. 8 ft
B. $8\frac{1}{2}$ ft
C. $7\frac{5}{6}$ ft
D. $8\frac{1}{6}$ ft

6. Aysha's mom gave her $6.24 to split with her brother at the mall. Use the base-ten blocks to determine how much money they each get.

A. $3.22
B. $3.12
C. $3.24
D. $2.12

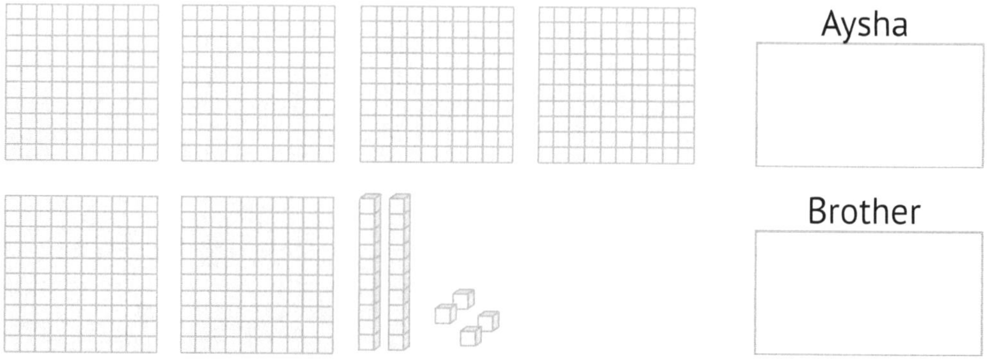

7. Solve.

$$\begin{array}{r} 681 \\ \times\ 23 \\ \hline \end{array}$$

A. 15,663
B. 14,663
C. 15,673
D. 15,763

8. Will made a Lego wall that is $\frac{2}{3}$ foot wide and $\frac{4}{5}$ foot long. What is the area of the Lego wall?

A. $\frac{6}{8}$ ft^2
B. $\frac{8}{15}$ ft^2
C. 1 ft^2
D. $1\frac{1}{2}$ ft^2

9. Bryant was at a birthday party. There was $\frac{7}{8}$ths of a pizza left. He at $\frac{3}{4}$ of the remaining pizza. How much of the pizza did he eat?

Use the figure to determine $\frac{3}{4}$ of $\frac{7}{8}$.

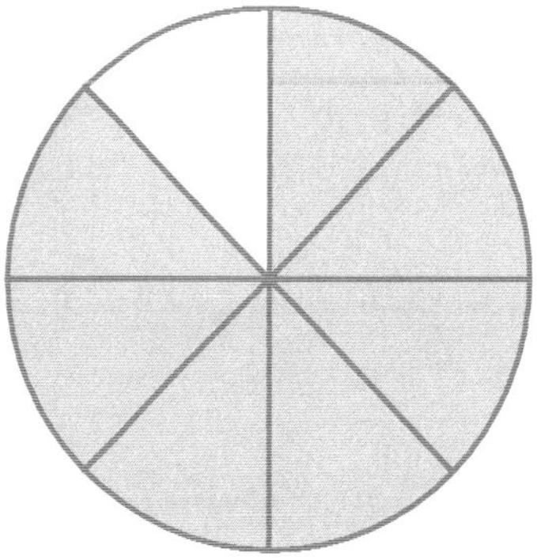

A. $\frac{10}{12}$
C. $\frac{10}{24}$
B. $\frac{21}{24}$
D. $\frac{21}{32}$

GO ON

10. The tooth fairy came to visit Daisy last night and left $2.22. How much money will Daisy get if she loses 5 teeth? Use the model below to show your work.

A. $10.11
B. $10.20
C. $11.10
D. $11.20

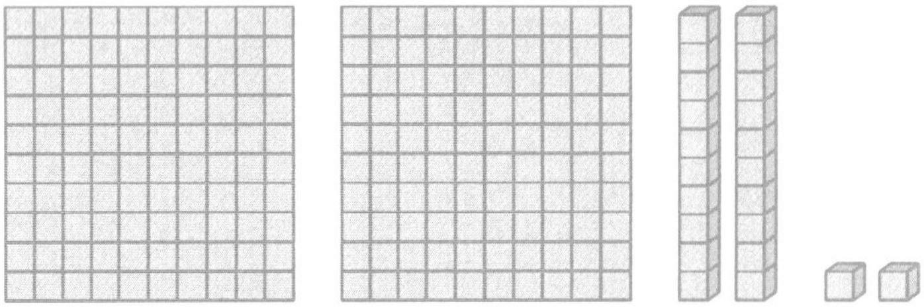

11. Which partial product would help fill in the area model?

A. 100
B. 1,000
C. 10,000
D. 700

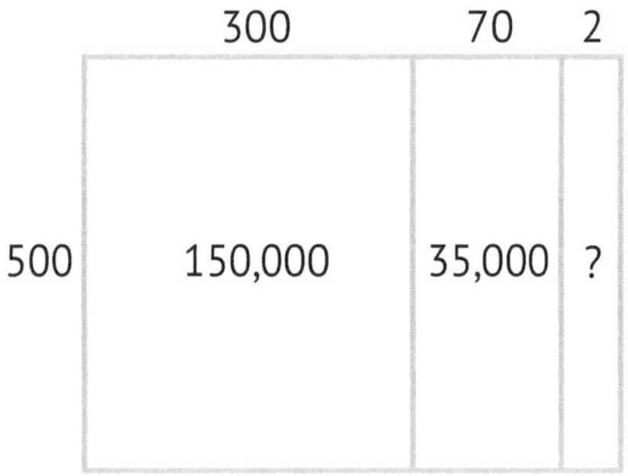

12. Graham is walking his dog. The leash is $3\frac{1}{2}$ feet long. When they get home his dog chews up $\frac{7}{8}$ ft of the leash. About how long is the leash now?

A. 2 ft　**B.** $2\frac{1}{2}$ ft　**C.** 3 ft　**D.** $3\frac{1}{4}$ ft

13. Anna is trying to figure out how many tiles to put on each wall. She has 1,842 tiles. She is decorating three walls and wants to put an equal number on each wall. Which equation will help her divide correctly?

A. 6 x 3 = 18

B. 60 x 3 = 180

C. 600 x 3 = 1,800

D. 6,000 x 3 = 18,000

14. There are 5 cakes left over from a neighborhood carnival. Six families want to split the cakes evenly. How much cake will each family get?

A. $\frac{5}{6}$ of a cake

B. $\frac{6}{5}$ of a cake

C. 1 cake

D. $\frac{1}{2}$ cake

15. Julie sold 5,000 cookies during a fundraiser. Ava sold 50 cookies. About how many fewer cookies did Ava sell compared to Julie?

A. 10 times as many

B. 100 times as many

C. $\frac{1}{10}$ as many

D. $\frac{1}{100}$ as many

GO ON

16. Wilson is trying to draw an isosceles triangle. What attributes will he need to include?

A. Two obtuse angles

B. An obtuse angle

C. All acute angles

D. All of the above

17. Oliver is building a treehouse. He draws up a plan for the house to be 12 feet long, 6 feet wide, and 5 feet tall. How many cubic feet will be in his treehouse?

A. 132 ft^3

B. 123 ft^2

C. 360 ft^3

D. 300 ft^2

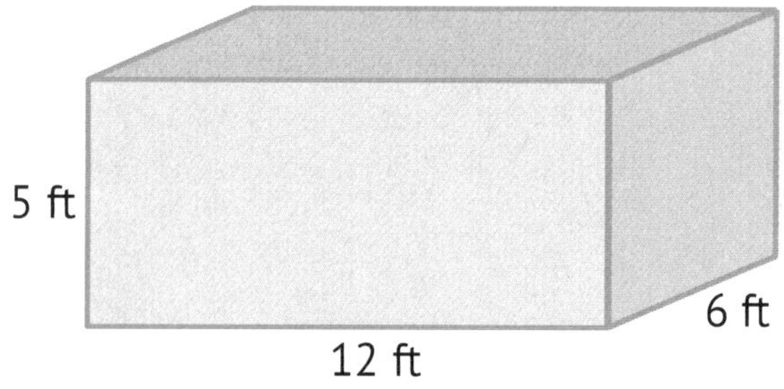

18. Francis works at the airport weighing luggage, but the scale is broken! It will only show ounces instead of pounds. The weight limit for a bag is 50 lbs. A passenger's bag weighed 896 ounces. Is it over or under the weight limit? If so, by how much?

1 lb = 16 oz

A. Under the limit by 6 lbs

B. Under the limit by 3 lbs

C. Over the limit by 6 lbs

D. Over the limit by 3 lbs

19. Which makes the expression true?

7.329 x ____ = 732.9

A. 100

B. 10

C. 1

D. 1,000

20. Connie is creating a border for a square bulletin board. Each side is 2.38 ft. How many feet of the border does she need to complete her project?

A. 9 ft

B. 9.64 ft

C. 9.52 ft

D. 10 ft

21. Margo is sampling licorice at a candy shop. She sampled 6 different pieces of licorice. Each one was $\frac{4}{5}$ of an inch long. How much licorice did Margo eat altogether? Use the model to multiply.

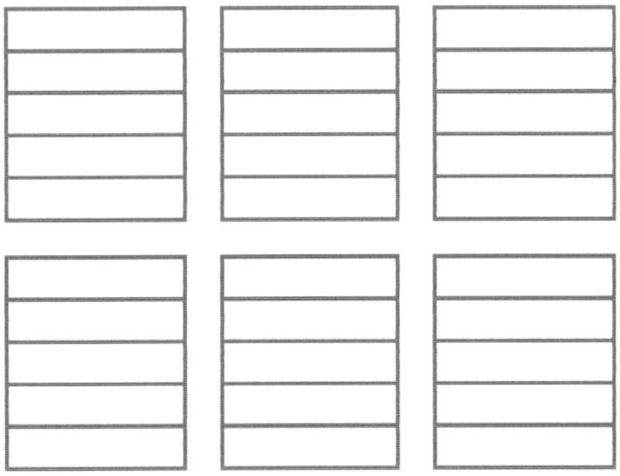

A. 5 inches

B. $4\frac{4}{5}$ inches

C. $4\frac{1}{2}$ inches

D. $4\frac{24}{30}$ inches

GO ON

22. Put the following expressions in order from least to greatest.

$$4 \times \frac{6}{6} \qquad 4 \times \frac{1}{5} \qquad 4 \times \frac{3}{2} \qquad 4 \times \frac{2}{3}$$

A. $4 \times \frac{3}{2}; 4 \times \frac{2}{3}; 4 \times \frac{1}{5}; 4 \times \frac{6}{6}$

B. $4 \times \frac{1}{5}; 4 \times \frac{2}{3}; 4 \times \frac{6}{6}; 4 \times \frac{3}{2}$

C. $4 \times \frac{2}{3}; 4 \times \frac{1}{5}; 4 \times \frac{6}{6}; 4 \times \frac{3}{2}$

D. $4 \times \frac{1}{5}; 4 \times \frac{2}{3}; 4 \times \frac{3}{2}; 4 \times \frac{6}{6}$

23. Damien wants to combine three large boxes to create a bookshelf in his room. Each box is 3 feet wide, 2 feet deep, and 2 feet tall. What will the volume of his bookshelf be?

A. 12 ft^2

B. 10 ft^3

C. 36 ft^2

D. 36 ft^3

24. George is working in a butcher shop. He is weighing ground beef. The scale says 2.189 lbs. What is the weight rounded to the nearest hundredths place?

A. 2.18 lbs

B. 2.20 lbs

C. 2.19 lbs

D. 2.00 lbs

25. Eric has $7.41 to buy concessions at a baseball game. He spends $1.25 on a soda and $2.50 on a hot dog. How much money does he have leftover? Use the model to show your work.

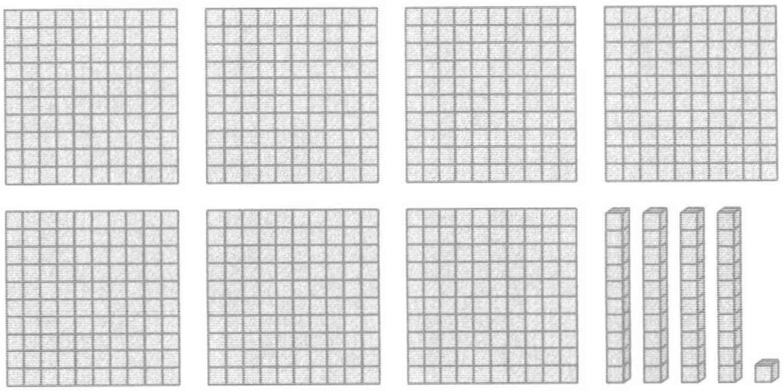

A. $3.66

B. $3.60

C. $4.66

D. $4.50

26. How many cubic units make up the volume of the shape below?

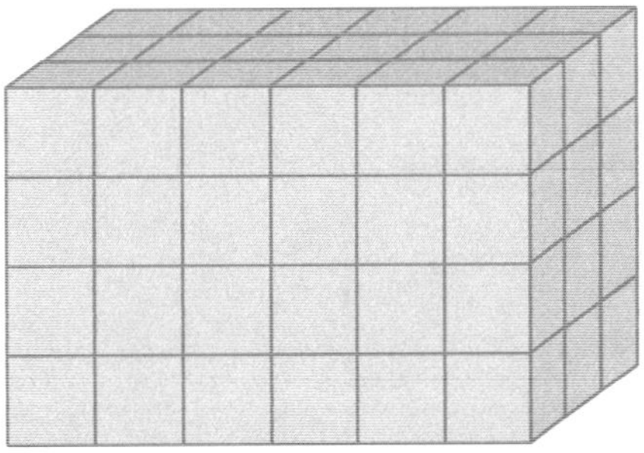

A. 54 cubic units

B. 72 cubic units

C. 80 cubic units

D. 52 cubic units

27. Anthony is cooking soup, but he can't find his measuring cup! He needs 3 cups of broth. How many tablespoons are in 3 cups of broth?

1 cup = 16 tablespoons

A. 32 tablespoons

B. 45 tablespoons

C. 48 tablespoons

D. 52 tablespoons

GO ON ▶

28. What is the difference between the longest and shortest runs done by the cross country club?

A. 1 mile

B. $7\frac{1}{2}$ miles

C. $\frac{3}{4}$ mile

D. 2 miles

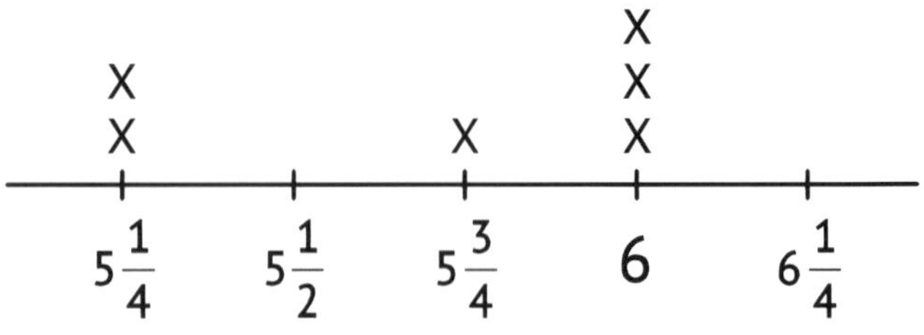

Miles Run by Cross Country Club

29. Which of the following is the equivalent standard form?

Two-hundred thirty-seven and twenty-four thousandths

A. 237.24
B. 237.024
C. 2,370.24
D. 237.240

30. Abby is giving out bubbles as a party favor. She made a big batch of bubble juice and wants to pour it into 3 oz bottles. She made 2 gallons of bubble juice. How many bottles can she fill?

1 gallon = 128 fl oz

A. 85 bottles, 1 oz leftover
B. 156 bottles
C. 42 bottles, 2 oz leftover
D. 384 bottles

Session 2

GO ON

31. Ricky listened to 4,281 minutes of music this year. If each song is about 3 minutes long, approximately how many songs did he listen to this year?

A. 1,723 songs
B. 1,147 songs
C. 1,427 songs
D. 1,327 songs

32. Zeke wants to create a line of toy cars that is longer than the couch. Each toy car is $\frac{3}{4}$ of an inch long. The couch is 48 inches long. How many toy cars would he need? (Hint: Work backwards.)

$$\_\_\_\_ \times \frac{3}{4} = \frac{}{4} = 48$$

A. 64 cars **B.** 50 cars **C.** 55 cars **D.** 58 cars

33. Matt and Leigh Anne are trying to find the volume of a rectangular box.

They both know they need to fill it with cubes, but their cubes are all different sizes. Leigh Ann says it doesn't matter that they are all different sizes, but Matt says they can't get the correct volume using the cubes. Who is correct? Explain your reasoning. How should they find the volume of the box?

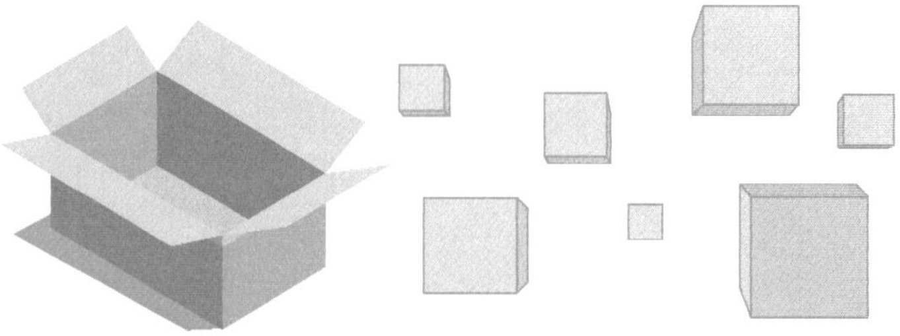

Explain your reasoning.

____

34. A toy factory in Bejing shipped 216 containers of rubix cubes. Each container had 24 rubix cubes. How many were shipped total?

A. 5,184 rubix cubes
B. 5,094 rubix cubes
C. 5,175 rubix cubes
D. 4,184 rubix cubes

35. Last month the Cub Scouts raised $416 selling popcorn. Their goal for their next fundraiser is to earn 100 times more. How much do they want to earn during the next fundraiser?

A. $4.16
B. $41.16
C. $4,160
D. $41,600

36. It has been raining all week. Stuart has been collecting and measuring rainfall amounts.

Monday	$\frac{3}{4}$ inch	Wednesday	$\frac{5}{6}$ inch
Tuesday	$2\frac{1}{2}$ inches	Thursday	$\frac{2}{12}$ inch

He says that his town has received more than 6 inches of rain. Is he correct? **Explain your reasoning and show your work.**

____

GO ON ▶

37. After a New Year's party, there is $\frac{1}{5}$ of a cake left. 5 friends want to divide it evenly. Use the model below to show how much cake each person would get.

A. $\frac{1}{30}$

B. $\frac{1}{5}$

C. $\frac{1}{25}$

D. $\frac{1}{10}$

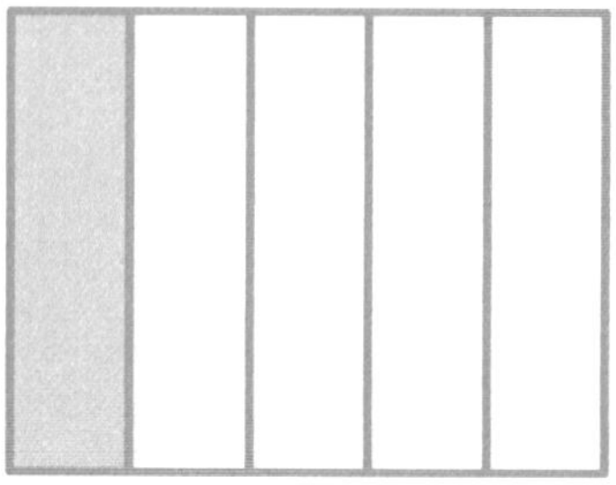

38. The cheerleading team is creating a banner for the homecoming rally. They want to create a checkerboard with the school colors. Each square is 2 inches wide. They want the banner to be 15 yards wide. How many squares will be featured on the width of the banner?

12 inches = 1 foot

3 feet = 1 yard

A. 270 squares

B. 210 squares

C. 54 squares

D. 48 squares

39. Use the data below to complete the line plot.

$3\frac{1}{4}$ inches, 2 inches, $2\frac{1}{2}$ inches, 3 inches, $3\frac{1}{4}$ inches, 2 inches, 2 inches

Montly Rainfall (Inches)

40. Evaluate the expression:

$$\left(\frac{32}{8}\right) + \frac{(12 \times 2)}{6} = \_\_\_\_\_\_$$

A. 11
B. 8
C. 12
D. 7

41. Robert is sure that this class is lasting forever. He estimates that he's been sitting in class for more than 10,000 seconds. The class is 2 hours long. How many seconds is the class?

1 hr = 60 minutes

1 minute = 60 seconds

A. 10,000 seconds
B. 7,200 seconds
C. 3,600 seconds
D. 1,200 seconds

42. Cassie says that 89.25 x 100 would be 8,900.25. Is she correct or incorrect?

Explain your reasoning.

____

____

GO ON ▶

43. Shawn solved the math problem below.

$$\begin{array}{r} \overset{1}{5}73 \\ \times \quad 21 \\ \hline \overset{1}{5}73 \\ +\ 1146 \\ \hline 1,719 \end{array}$$

Is he correct or incorrect? Explain how he got this answer.

____

44. A waiter gives a table 6 rolls in a basket. There are 7 people at the table. If the rolls are divided evenly, how much will each person get?

A. $\frac{7}{6}$ of a roll

B. $\frac{1}{42}$ of a roll

C. $\frac{1}{13}$ of a roll

D. $\frac{6}{7}$ of a roll

45. Olivia needs to find (5,13) on the coordinate plane, but she isn't sure how. Explain to her how she can find and label the point on the coordinate plane.

____

____

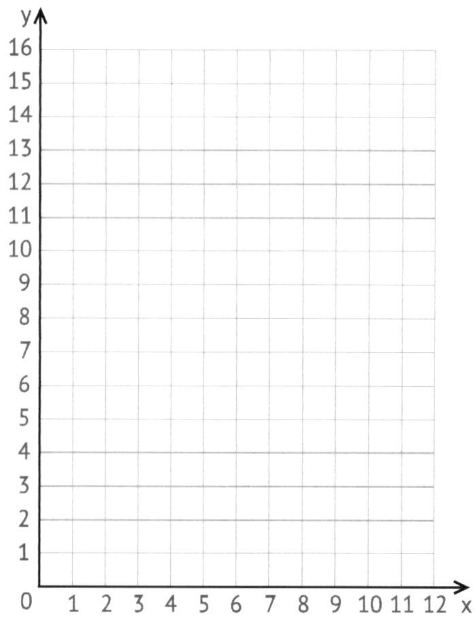

Test 3: Session 1

Tips: Be sure to read each question carefully and work through the problem step by step before choosing the correct answer.

Suggested Time:

Session 1: 55 to 65 Minutes
Session 2: 60 to 75 minutes

Students are **NOT** permitted to use a calculator.

1. Oscar is sending blankets overseas to soldiers. He has 212 blankets. He has 7 boxes and he wants to make sure each box has an equal number of blankets. Which equation could he use to determine about how many blankets would fit in each box?

A. 7 x 3 = 21

B. 7 x 30 = 210

C. 7 x 300 = 2,100

D. 7 x 3,000= 21,000

2. Portsmouth, New Hampshire got 3 feet of snow last night. How many inches of snow did they get?

A. 30 inches

B. 36 inches

C. 27 inches

D. 18 inches

3. Gary is making a vegetable stew. He adds $\frac{3}{4}$ cups of carrots, $1\frac{1}{2}$ cups of potatoes, and $\frac{2}{3}$ cups of celery. He thinks he will have around 2 cups of vegetables. About how much will be in the stew?

A. Less than 1 cup

B. Around $1\frac{1}{2}$ cups

C. Around 2 cups

D. Around 3 cups

4. Matt and Michael ran a 100-meter race. Matt finished in 17.214 seconds. Michael finished in 17.241 seconds. Who finished first?

A. Matt

B. Michael

C. It was a tie

D. All of the above

5. Students logged their time spent reading during a reading marathon. What is the difference between the longest and shortest time spent reading?

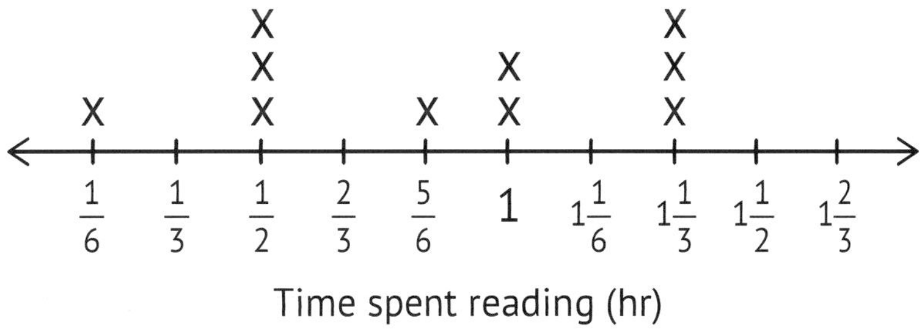

A. $1\frac{1}{3}$ hours **B.** $1\frac{2}{3}$ hours **C.** $1\frac{1}{6}$ hours **D.** 1 hour

6. What is the total amount of time the class spent reading?

A. 8 hours **B.** $8\frac{1}{2}$ hours **C.** $8\frac{1}{6}$ hours **D.** $7\frac{1}{2}$ hours

GO ON

7. Look at the line plot in question 5. What is the median amount of time spent reading?

A. 1 hour

B. $\frac{1}{2}$ hour

C. $\frac{5}{6}$ hour

D. 10 hours

8. A new pickup truck weighs 2.5 tons. How many pounds does it weigh?

A. 2,000 lbs
B. 2,500 lbs
C. 5,000 lbs
D. 5,500 lbs

9. A laundry basket is 3 feet long, 2 feet wide, and 1 foot tall. What is the volume of the laundry basket in cubic inches?

A. 6 ft^3
B. 5ft^3
C. 6ft^2
D. 5ft^2

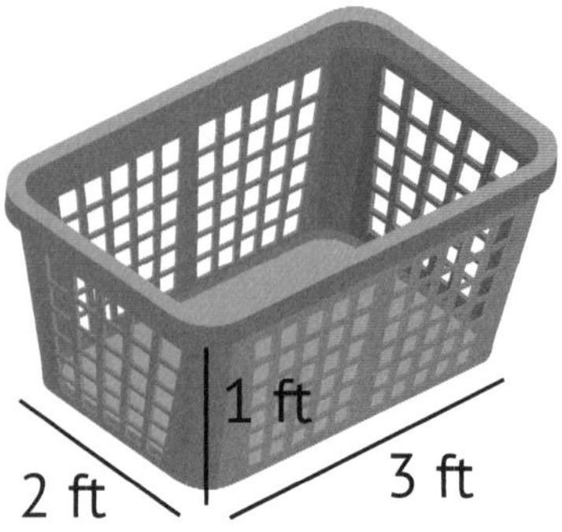

10. Pam and Jim are trying to buy tickets at a carnival. Pam has $2.41 and Jim has $3.58. How much do they have altogether? Use the base ten blocks to determine how much money they have to buy tickets.

A. $5.99
B. $6.01
C. $6.00
D. $4.99

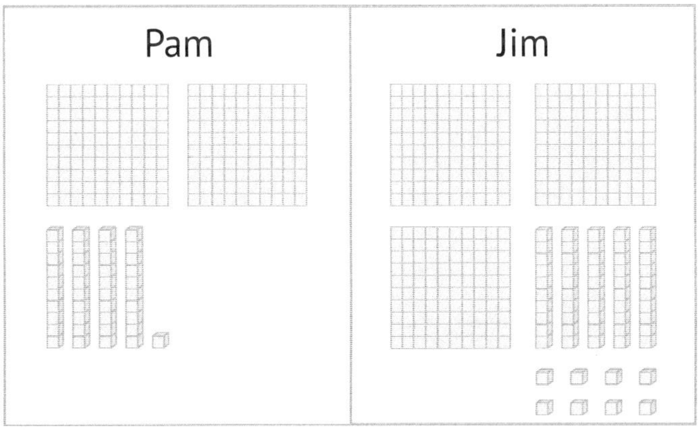

11. Leslie is preparing food for her dogs. She mixes $3\frac{1}{4}$ cups of dry food and $2\frac{2}{3}$ cups of wet food. She adds $\frac{1}{2}$ cup of water and stirs. How many cups of ingredients does she use for her dogs' food in all?

A. $5\frac{4}{9}$ cups **B.** $6\frac{5}{12}$ cups **C.** $5\frac{5}{12}$ cups **D.** $6\frac{1}{2}$ cups

12. Sarah went to the coffee shop 4 days last week. Each time she had $2\frac{1}{2}$ cups of coffee. How much coffee did she drink last week?

A. 10 cups
B. 5 cups
C. 20 cups
D. $8\frac{1}{2}$ cups

GO ON ▶

13. Eric just can't stop getting speeding tickets! He's gotten 15 every month. How many speeding tickets did he get this year?

A. 150

B. 250

C. 180

D. 280

14. Marcus is trying to buy a garden shed. He needs it to be at least 1,000 cubic feet. Which of the following measurements would be large enough?

A. 10 ft long, 10 ft wide, 12 ft tall

B. 9 ft long, 8 ft wide, 5 ft tall

C. 7 ft long, 8 ft wide, 5 ft tall

D. 10 ft long, 9 ft wide, 6 ft tall

15. Last year a paper sales department made $45,192 in sales. This year they made $\frac{1}{100}$ the amount from last year. How much money did they make this year?

A. $4,519,250

B. $451.92

C. $45.19

D. $45,192,500

16. Andrew is distributing post-its to his coworkers. He got a pack of 250 post-its. There are 9 people who work in his office. He is using an area model to divide. Which partial quotient would help him solve?

A. 6

B. 7

C. 8

D. 9

	20	?	
9	250 - 180 70	70	

17. There is $\frac{1}{4}$ of a cake leftover. Four friends want to split it evenly. How much cake will each get?

A. $\frac{1}{8}$

B. $\frac{1}{16}$

C. $\frac{1}{12}$

D. $\frac{1}{4}$

18. Which expression matches the phrase "the sum of 4 and 20 divided by 6?"

A. $4 + \frac{20}{6}$

B. $4 \times \frac{20}{6}$

C. $\frac{(4 + 20)}{6}$

D. $4 - \frac{20}{6}$

19. Julie wants to split her $\frac{1}{2}$ gallon of milk among 5 students. How much milk will each student get?

A. 10 gallons **B.** $\frac{1}{2}$ gallon **C.** $\frac{1}{10}$ gallon **D.** $\frac{1}{7}$ gallon

20. Rosa is analyzing a data set. She determines that the rule is "add 5."

x	y	(x,y)
0	5	(0,5)
1	6	(1,6)
2	7	(2,7)
3	__	(3,__)

What is the missing data point?

A. 8 **B.** 9 **C.** 10 **D.** 15

21. Elizabeth is trying to fill a container with cereal. Which would be the best unit to help her determine the volume of the container?

A. Centimeters
B. Meters
C. Kilometers
D. Yards

22. Kevin has 527 chickens on his farm. He has to sell the farm and he will be moving the chickens to three new farms. He is trying to determine how many chickens to send to each farm using the partial quotient method.

$$\begin{array}{r|r} & 100 \\ \hline 3 & 527 \\ & -300 \\ \hline & 227 \end{array}$$

What will his next partial quotient be?

A. 7 **B.** 70 **C.** 700 **D.** 60

23. Solve the equation:

$$\mathbf{4.591 \times 10^2} = \_\_\_\_\_\_\_\_$$

A. 4.591
B. 45.91
C. 459.1
D. 4,591

24. Kevin is measuring the weight of penguins in the zoo. The penguin he is measuring weighs 5.719 lbs. What is the penguin's weight to the nearest tenth?

A. 5 lbs
B. 10 lbs
C. 5.7 lbs
D. 5.72 lbs

GO ON

25. A group of 23 volunteers went to canvass before a local election. They knocked on 31 doors each. How many doors did they knock on altogether?

A. 713 **B.** 230 **C.** 210 **D.** 720

26. Meredith's grandmother always sends her money in her birthday card. She has been saving the money for nine years. Each year her grandmother sends $4.25. How much money does Meredith have on her 9th birthday? Draw base ten blocks to determine the amount.

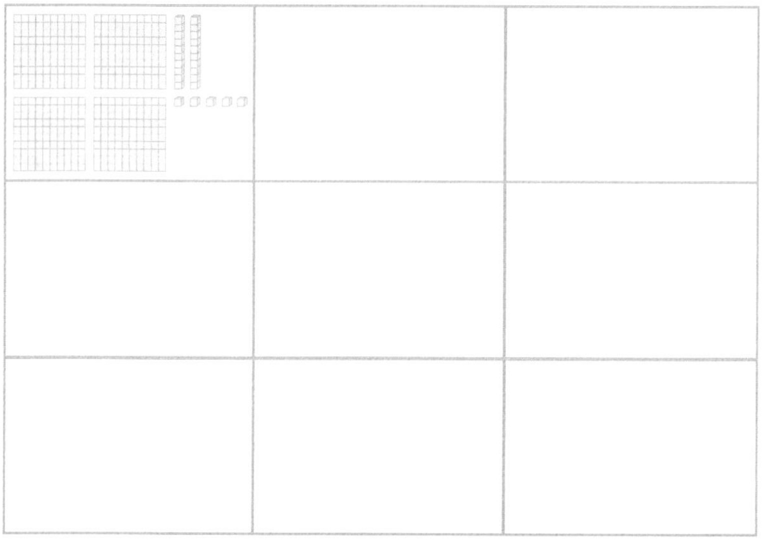

A. $39.00
B. $38.75
C. $38.50
D. $38.25

27. Donna is mixing gallons of paint colors. She adds the following amounts:

How much paint will she have after she mixes them all together?

A. 1 gallon **B.** $1\frac{1}{2}$ gallon **C.** $\frac{9}{10}$ gallon **D.** $1\frac{1}{9}$ gallon

28. A group of six cats finds 5 pieces of cheese. They split the cheese evenly. How much cheese do each of them get?

A. $\frac{6}{5}$ of a piece

B. $\frac{5}{6}$ of a piece

C. 30 pieces

D. 1 piece

29. April has saved up candy bars from Halloween. She eats $\frac{1}{4}$ of a candy bar every day. She has been enjoying them for 9 days. After 9 days, how much of the candy bars has she eaten?

A. 2 candy bars

B. $2\frac{1}{2}$ candy bars

C. $2\frac{1}{4}$ candy bars

D. $1\frac{3}{4}$ candy bars

30. Estimate the size of the product:

$$42 \times \frac{5}{3} = \_\_\_\_\_\_$$

A. Greater than 42
B. Equal to 42
C. Less than 42
D. Equal to 0

GO ON

Session 2

31. A town is experiencing a blizzard. On Thursday they received $1\frac{5}{6}$ feet of snow. On Friday they received $1\frac{2}{3}$ feet of snow. How much snow did they receive in all?

A. 3 feet
B. $2\frac{5}{6}$ feet
C. $2\frac{2}{3}$ feet
D. $3\frac{1}{2}$ feet

32. Claire is weighing gifts to mail them. She knows they weigh 12 pounds, but the post office charges by the ounce. How many ounces are in 12 pounds?

A. 24 ounces
B. 144 ounces
C. 192 ounces
D. 240 ounces

33. Find the missing points in the data set and determine the rule.

x	y	(x, y)
24	12	(24, 12)
20	10	(20, 10)
___	8	(___, 8)
12	___	(12, ___)
Rule = ______________________		

Explain how you were able to determine the rule.

____

GO ON ▶

34. Adam is tracking the growth of his tomato plant. It grows $\frac{2}{5}$ cm every day. He tracks it for seven days. How much has it grown after seven days?

A. 2 cm

B. $2\frac{1}{4}$ cm

C. $2\frac{4}{5}$ cm

D. $2\frac{1}{5}$ cm

35. Compare the decimals using <, >, or =.

$$\mathbf{48.71} \bigcirc \mathbf{48.712}$$

Explain the process you used to determine the value of the decimals. How did you compare them?

____

____

36. Laura is working at a local food pantry to support needy families in her community. She is trying to calculate how much food can fit in a box. What measurements does she need to take? How will she use those measurements to determine the capacity of each box? Why would this be helpful information for her?

Explain your reasoning.

____

____

37. Eve is arranging snacks on a tray. She notices that $\frac{3}{4}$ of the 12 snacks are vegetables, so she puts them into four equal sized groups. Use the visual model to show $\frac{3}{4}$ of 12.

There are ________ vegetables on the snack tray.

38. A teacher is trying to determine how many cups of hot chocolate he can give to his students. His recipe yielded 3 gallons. Which operation - multiplication or division - will help him determine the number of cups? Why?

____

____

39. Classify the triangle by describing its angles and sides.

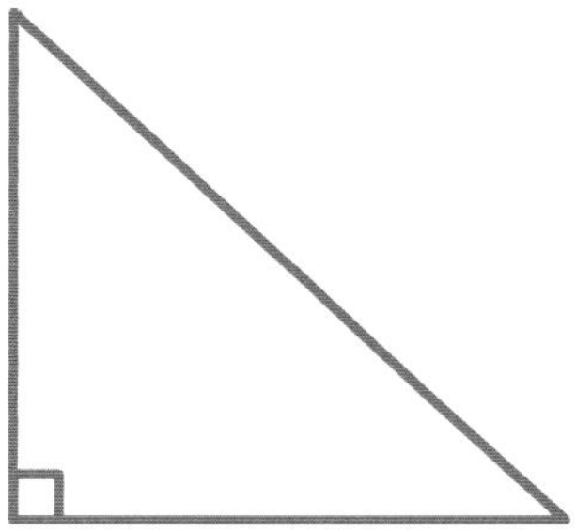

____

____

____

GO ON

40. Once a month Stuart makes a $8.65 donation to the library. How much does he donate in a year?

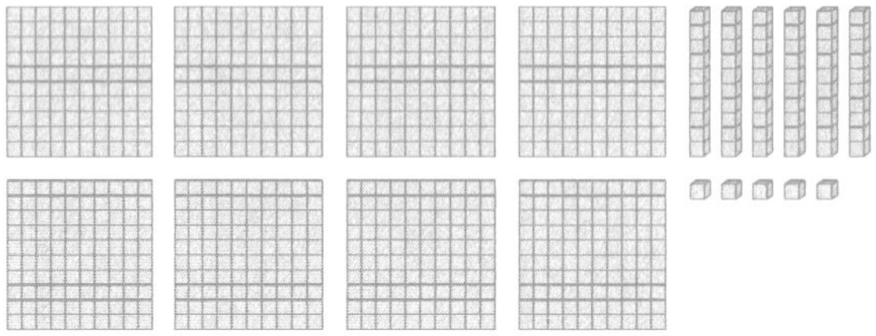

A. $96.45 **B.** $225.96 **C.** $103.80 **D.** $98.50

41. What is the missing partial quotient?

```
      8
     [ ]
    300
4 | 1432
  - 1200
     232
   - 200
      32
    - 32
       0
```

A. 5 **B.** 50 **C.** 6 **D.** 60

42. Which equation would yield the product 1,615?

A. 32 x 29
B. 85 x 19
C. 22 x 81
D. 61 x 12

43. What is 943.219 rounded to the nearest hundredth?

A. 900

B. 1,000

C. 943.22

D. 943.2

44. Monica is solving a multiplication problem using the traditional algorithm. She multiplies 49 times 24 and gets the product 1,149. What was her error? Show her the correct way to solve.

____

____

45. Dylan is working on solving the following expression:

$$4 + \frac{(3 \times 12)}{6} = \_\_\_\_$$

He says, "I added 4 plus 3 to get 7. Then 7 times 12 is 84. 84 divided by 6 is 14."

What was his error? How can he solve it correctly? What is the actual value of the expression?

____

____

GO ON

Test 4: Session 1

Tips: Be sure to read each question carefully and work through the problem step by step before choosing the correct answer.

Suggested Time:

Session 1: 55 to 65 Minutes
Session 2: 60 to 75 minutes

Students are **NOT** permitted to use a calculator.

1. A barge carrying shipping containers is traveling up the river. They are helping to transport new furniture from a factory to a store. There are 25 shipping containers. Each one has 18 sofas inside. How many sofas are being shipped?

A. 450
B. 430
C. 425
D. 460

2. What is the value of the decimal below? What is its value rounded to the nearest tenth?

A. 7.35; It rounds to 7.3
B. 7.36; It rounds to 7.4
C. 7.37; It rounds to 7.4
D. 7.38; It rounds to 7.3

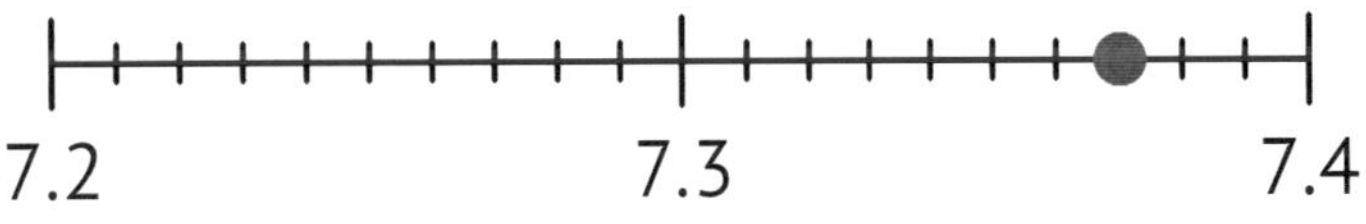

3. Oliver is making fruit punch for a huge party. He has made 3 gallons of fruit punch. Each drink dispenser can hold $\frac{1}{2}$ gallon. How many drink dispensers can he fill?

A. 3
B. 4
C. 5
D. 6

GO ON ▶

4. The rectangle is one square unit:

Which equation matches the figure?

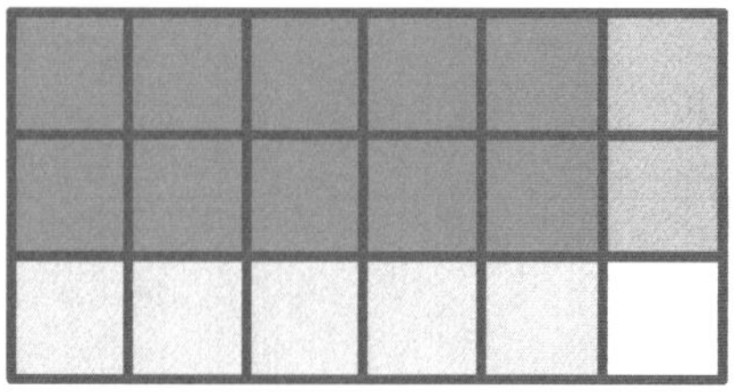

A. $\frac{1}{3} \times \frac{5}{6}$

B. $\frac{2}{3} \times \frac{5}{6}$

C. $\frac{1}{4} \times \frac{4}{6}$

D. $\frac{2}{3} \times \frac{4}{6}$

5. Maggie made Valentine's gift boxes for her classmates. She stacked them up while looking for a container. How many Valentine's boxes can fit her container?

A. 18
B. 9
C. 27
D. 21

6. Ben bought some plant food for his massive garden. He mixed it with water and it yielded 5 gallons of plant food. Each raised flower bed needs $\frac{1}{3}$ of a gallon. How many flower beds can he feed?

A. 10
B. 15
C. 20
D. 25

7. Aubrey is studying for exams. On Monday she studies $\frac{5}{6}$ of an hour. On Tuesday she studies for $2\frac{2}{3}$ hours. On Wednesday she studies for $\frac{3}{4}$ of an hour. She thinks she studied for five hours this week. About how many hours did she study?

A. About 2 hours

B. About 3 hours

C. About 4 hours

D. About 5 hours

8. How many people ran the 100-meter race in 18.75 seconds?

A. 3

B. 4

C. 5

D. 6

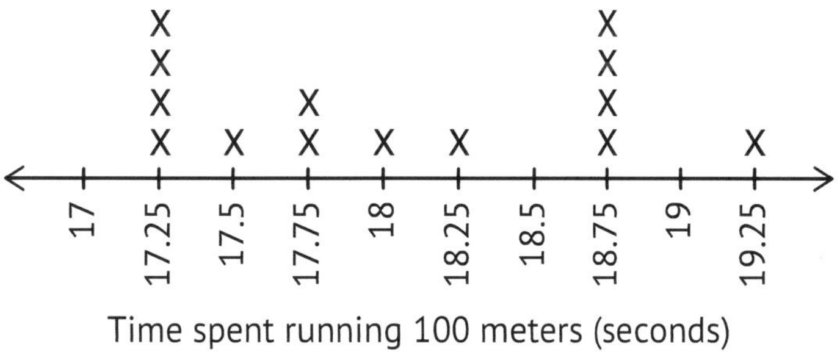

9. Stanley and his friends are putting together Easter baskets for a neighborhood celebration. They have collected 31 baskets. They want to put 15 eggs in each one. How many eggs do they need?

A. 460

B. 455

C. 465

D. 450

GO ON

10. A local hockey team is hosting a baby race as a fundraiser. Babies must crawl a total of 20 yards in the race. How many feet must the babies crawl?

A. 32 feet
B. 240 feet
C. 60 feet
D. 40 feet

11. Edna just turned 95 years old. She had 9 friends bring her a birthday cake to celebrate, and each of them had 95 candles! How many birthday candles does Edna have to blow out?

A. 180
B. 900
C. 864
D. 855

12. Which expression is greater than 724?

A. 724×1 **B.** $724 \times \frac{6}{2}$ **C.** $724 \times \frac{4}{4}$ **D.** $724 \times \frac{1}{3}$

13. Jerry has a huge yard. It takes him several days to mow the whole thing. On Friday he mows $\frac{4}{12}$ of the yard. On Saturday he mows $\frac{1}{4}$ of the yard. Has he mowed the whole yard? How much is left?

A. He mowed the entire yard.

B. He is not done yet. He has less than $\frac{1}{4}$ left to mow.

C. He is not done yet. He has more than $\frac{1}{4}$ left to mow.

D. He is not done yet. He has more than $\frac{1}{2}$ left to mow.

14. The art teacher is running out of paint! There is $\frac{1}{2}$ cup of red paint left. There are 6 students that want to use it. How much would each student get if she splits it evenly?

A. $\frac{1}{2}$ cup

B. $\frac{1}{8}$ cup

C. $\frac{1}{12}$ cup

D. 12 cups

15. There are 12 monkeys in a zoo. The zookeeper only has 11 bananas. He wants to split the bananas equally between the monkeys. How much banana will each monkey get?

A. 132 pieces

B. 1 piece

C. $\frac{12}{11}$ of a banana

D. $\frac{11}{12}$ of a banana

GO ON

16. Toby had $7.31 in his wallet. He spent $1.25 on a candy bar and $2.35 on a soda. How much money does he have left? Use the base ten blocks to determine the amount he has leftover.

A. $4.71
B. $3.75
C. $3.71
D. $4.00

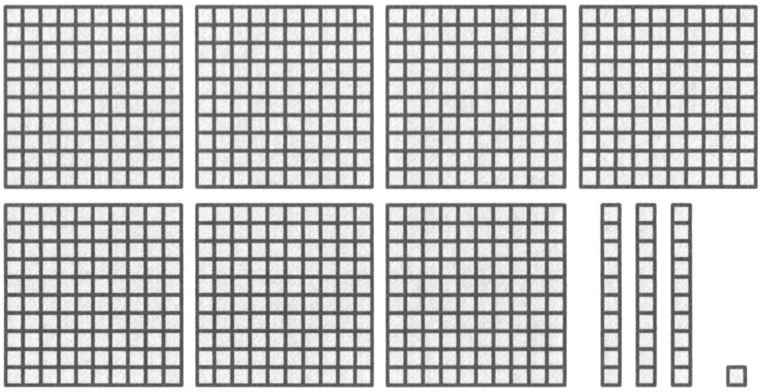

17. Mark is making 10 cakes at his bakery. Each cake needs $3\frac{2}{3}$ cups of frosting. How much frosting does he need for all 10 cakes?

A. 30 cups
B. $36\frac{2}{3}$ cups
C. $37\frac{1}{2}$
D. 32 cups

18. Craig brought leftover cake into the office. When he brought it in there were $\frac{8}{10}$ pieces. On Monday his coworkers at $\frac{3}{10}$. On Tuesday they at $\frac{2}{10}$. How many pieces were left on Wednesday?

A. $\frac{2}{10}$
B. $\frac{3}{10}$
C. $\frac{4}{10}$
D. $\frac{5}{10}$

19. Greg is building a block tower. Each block is $2\frac{1}{2}$ inches tall. He has used 12 blocks. How tall is his block tower?

A. 60 inches

B. 62 inches

C. 30 inches

D. 32 inches

20. Bernard is filling a storage box with Legos. Which would be the best unit to measure the volume of the storage box?

A. Inches

B. Yards

C. Kilometers

D. Meters

21. Eric is packing up his library. Each box can hold 5 lbs. How many ounces can each box hold?

A. 50 ounces

B. 80 ounces

C. 25 ounces

D. 250 ounces

GO ON

22. What does volume measure?

A. The outside of a shape
B. The inside of a shape
C. The surface area of a shape
D. The weight of a shape

23. A Scout leader is distributing stickers for scouts to use on their Pinewood Derby cars. The Scout leader has 391 stickers. There are 8 children in the scout troop. Use the area model to determine how many stickers each child will receive.

A. 487
B. 47 with 6 leftover
C. 48 with 7 leftover
D. 46 with 6 leftover

8	391		

24. A toy company made approximately 65,000 toys last year. This year they made 650,000 toys. Which comparative statement shows the relationship between last year's sales and this year's sales?

A. This year they made $\frac{1}{100}$ the number of toys compared to last year.
B. This year they made $\frac{1}{10}$ the number of toys compared to last year.
C. This year they made 10 times the number of toys compared to last year.
D. This year they made 100 times the number of toys compared to last year.

25. What is the coordinate pair for the point on the plane below?

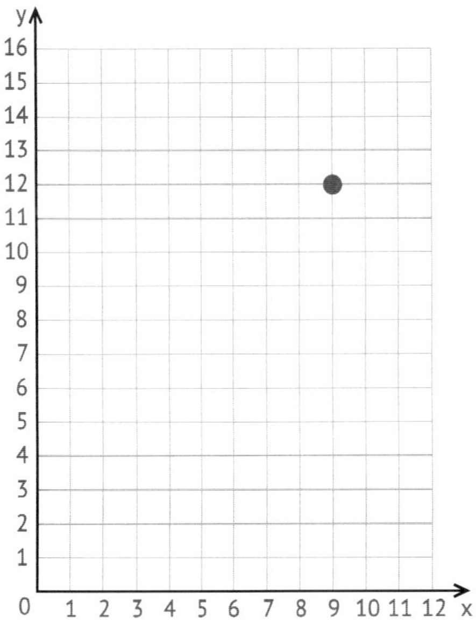

A. (9, 12)

B. (12, 9)

C. 12,9

D. 9,12

26. Angela is stocking shelves with cans of soup in her store. There are 212 cans of soup and 5 shelves. Which equation will help her make the best estimate of the number of cans per shelf?

A. 5 x 4 = 20

B. 5 x 40 = 200

C. 5 x 5 = 25

D. 5 x 50 = 250

27. Ron is doing a woodworking project. He had a piece of wood that was $10\frac{7}{8}$ feet long. He cuts off a piece that is $3\frac{3}{4}$ feet long. How much of the wood is left?

A. $7\frac{1}{8}$ feet

B. $7\frac{1}{4}$ feet

C. 7 feet

D. $8\frac{1}{8}$ feet

GO ON

28. Paul is buying moving boxes. He purchases boxes that are 3 feet long, 3 feet wide, and 4 feet tall. How many cubic feet are the moving boxes?

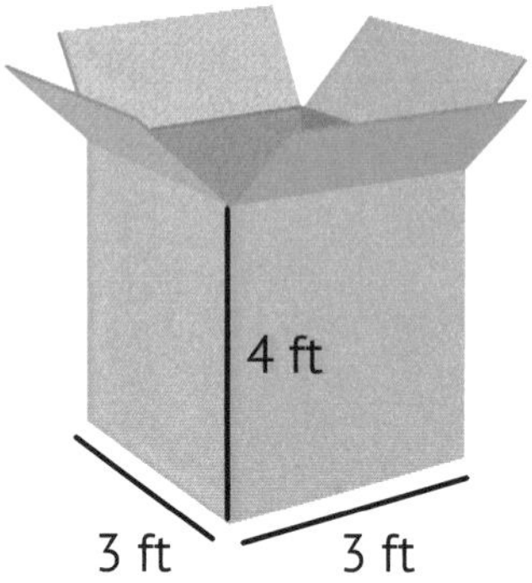

A. 10 ft^2

B. 10 ft^3

C. 36 ft^2

D. 36 ft^3

29. Daryl is gluing shoeboxes together to make a birdhouse. The first box is 12 inches long, 6 inches wide, and 4 inches high. The second box is 6 inches long, 5 inches wide, and 5 inches high. What is the combined volume of the two boxes?

A. 430 inches

B. 438 inches

C. 440 inches

D. 288 inches

30. During his marathon training, David ran 4,521 meters. How many kilometers did he run?

A. 4.521 kilometers

B. 45.21 kilometers

C. 452.1 kilometers

D. 45,210 kilometers

Session 2

GO ON

31. Use the coordinate pairs to plot points on the coordinate plane.

(0,6) (2,1) (3,4) (5,2) (7,7)

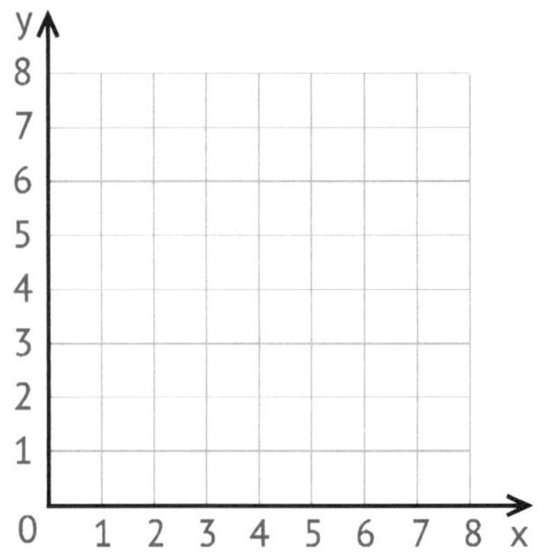

32. A scientist discovers a new planet. It's radius is $\frac{1}{100}$th the size of Earth's radius. Earth's radius is 3,958.8 miles across. What is the length of the new planet's radius?

A. 39,588 miles

B. 395.88 miles

C. 39.588 miles

D. 3.9588 miles

33. Frank is measuring the length of mice in a science lab. Use the data to create a line plot.

Length of mice in inches: $\mathbf{3\frac{1}{4}, 2\frac{3}{4}, 3, 3\frac{1}{2}, 3\frac{1}{4}, 2\frac{1}{2}, 2\frac{3}{4}, 1\frac{1}{2}, 2\frac{1}{4}, 3\frac{1}{4}, 4, 3\frac{1}{2}, 3\frac{1}{4}, 2\frac{1}{2},}$

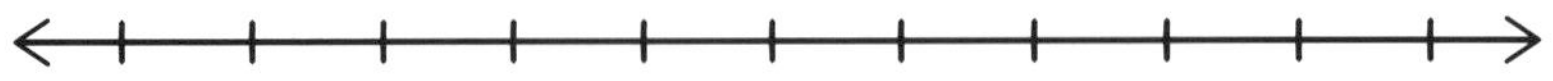

34. Joseph is solving a multiplication equation using the traditional algorithm. He multiplied 592 times 16 and got the product of 8,472. What was his error?

Show your work and explain your reasoning.

35. Find the missing measurement for the given volume.

Volume = 168 ft^3

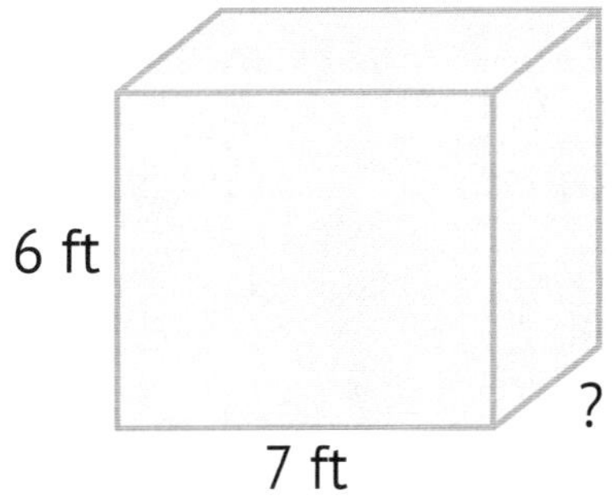

A. 3 ft
B. 4 ft
C. 5 ft
D. 6 ft

36. What is the standard form of the following decimal?

30,000 + 600 + 20 + 8 + .04 + .009

A. 30,000,600,208.04009
B. 36,028.049
C. 30,628.49
D. 30,628.049

GO ON

37. Use dotted lines to illustrate how to change the denominator of a fraction to add. Write the equation that matches the illustration and solve.

____

____

____

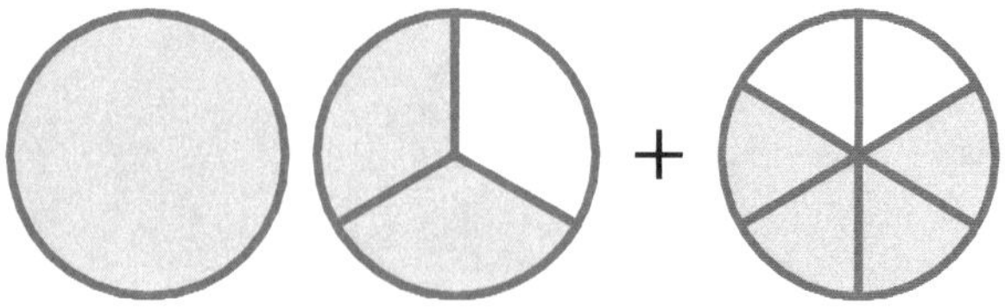

38. Ace practices music for $1\frac{5}{6}$ hours every day. He spends $\frac{2}{3}$ as much time on his homework. How much time does he spend on his homework?

A. 1 hour

B. $1\frac{2}{9}$ hours

C. $1\frac{3}{7}$ hours

D. $1\frac{7}{9}$ hours

39. Willow is solving an expression. It says "the product of 6 and 8 minus the quotient of 15 and 5." She says the answer is 6 remainder 3. What was her mistake? How should she write and solve the expression?

Show your work.

____

____

____

40. Circle all statements that correctly describe the shape.

A. It is a parallelogram.

B. It is a quadrilateral.

C. It is a square.

D. It is a rectangle.

Explain how it can be classified as more than one thing.

____

41. Margo is sewing together panels of fabric to make curtains. The first panel is $3\frac{3}{4}$ yards. The second panel is $6\frac{1}{2}$ yards. How many yards of fabric will she have when she combines them?

A. 10 yards

B. $10\frac{1}{4}$ yards

C. $10\frac{1}{2}$ yards

D. $10\frac{3}{4}$ yards

42. Eddie is trying to figure out how many meters are in 682.41 kilometers. He says there are .68241 meters. Is he correct? Why or why not?

____

____

____

____

43. Owen is making fruit punch for a party. He combines $\frac{3}{4}$ gallon of soda, $1\frac{1}{4}$ gallons of orange juice, and $\frac{1}{6}$ gallon of cranberry juice. How much will his recipe yield? Will he have more or less than 2 gallons? Use the visual models below to show the different ingredients.

He will make ______ gallons.

44. Solve using the area model.

A. 245
B. 250
C. 225
D. 230

3	735		

45. Solve

724 x 12 = ______.

A. 8,688
B. 8,868
C. 2,173
D. 2,172

Test 5: Session 1

Tips: Be sure to read each question carefully and work through the problem step by step before choosing the correct answer.

Suggested Time:

Session 1: 55 to 65 Minutes
Session 2: 60 to 75 minutes

Students are **NOT** permitted to use a calculator.

Camp counselors are buying sodas for the summer camp. The sodas are sold in packs of 12. There are 744 campers.

1. Which multiplication equation would help them estimate the number of packs to buy?

A. 12 x 40 = 480

B. 12 x 50 = 600

C. 12 x 60 = 720

D. 12 x 70 = 840

2. How many packs of soda do they need to buy?

A. 60

B. 62

C. 64

D. 66

3. Fred bought a leash that is 8 meters long. George bought a leash that is $\frac{4}{5}$ as long as Fred's. About how long is George's leash?

A. Between 3 and 4 meters

B. Between 4 and 5 meters

C. Between 5 and 6 meters

D. Between 6 and 7 meters

4. Kim is baking a cake. She needs 4 cups of flour, but she can only find her tablespoon. She will have to measure out 4 cups of flour using her tablespoon. How many tablespoons of flour does she need?

1 cup = 16 tbsp

A. 16 tbsp
B. 32 tbsp
C. 48 tbsp
D. 64 tbsp

5. Which number has a greater value than the expanded form?

(5 x 100) + (4 x 10) + (7 x 1) + (5 x .1) + (2 x .01) = ______

A. 547.53
B. 541.57
C. 547.51
D. 541.99

6. Which equation matches the figure below?

A. 5 x 4 = 20

B. $\frac{1}{5} \times \frac{2}{4} = \frac{2}{20}$

C. $\frac{1}{5} \times \frac{2}{4} = \frac{1}{20}$

D. $\frac{1}{5} \times \frac{2}{4} = \frac{16}{20}$

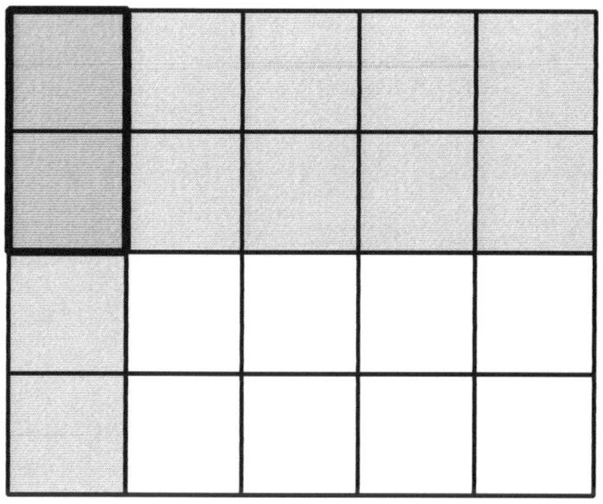

GO ON

7. There are 17 scheduled flights departing from the airport this morning. Each airplane can seat 416 travelers. How many people are departing from the airport this morning?

A. 7,078

B. 7,072

C. 7,182

D. 6,072

8. Paul wants to combine all of the dry ingredients for a cake into a bowl. He will use $1\frac{1}{2}$ cups of flour, $\frac{1}{5}$ cup of baking soda, $\frac{3}{10}$ cups of sugar, and $\frac{1}{10}$ cup of salt. What amount of dry ingredients does he need for his cake?

A. 2 Cups

B. $1\frac{7}{10}$ cups

C. $2\frac{1}{10}$ cups

D. $2\frac{2}{10}$ cups

9. Which dimension is not used when determining the volume of a solid figure?

A. Length

B. Height

C. Weight

D. Width

10. Oliver is generating a data set. So far he has the following numbers:

108, 99, 90, 81, 72...

What rule is helping him generate the data?

A. Subtract 9 each time
B. Divide by 9 each time
C. Subtract 11 each time
D. Divide by 11 each time

11. A dump truck can carry 22 cubic yards of mulch. The Parks Department orders 34 trucks to mulch the flower beds in parks throughout the city. How many cubic yards of mulch did the city order?

A. 740
B. 126
C. 748
D. 275

12. Which expression has the greatest value?

A. 425×10^2
B. 425×10^7
C. 425×10^3
D. 425×10^4

GO ON ▶

13. Six friends visit a lemonade stand. There are only 5 cups of lemonade left. They want to be fair and split the cups equally. How much will each friend get?

A. $\frac{5}{6}$ cup

B. 1 cup

C. $\frac{1}{30}$ cup

D. $\frac{1}{11}$ cup

14. In February Lacy sold a house worth $425,000. In March she sold a house that was one hundred times more expensive. How expensive was the house she sold in March?

A. $42,500

B. $4,250,000

C. $42,500,000

D. $425,000,000

15. Count the cubic units to determine the volume.

A. 27 cubic units

B. 9 cubic units

C. 12 cubic units

D. 6 cubic units

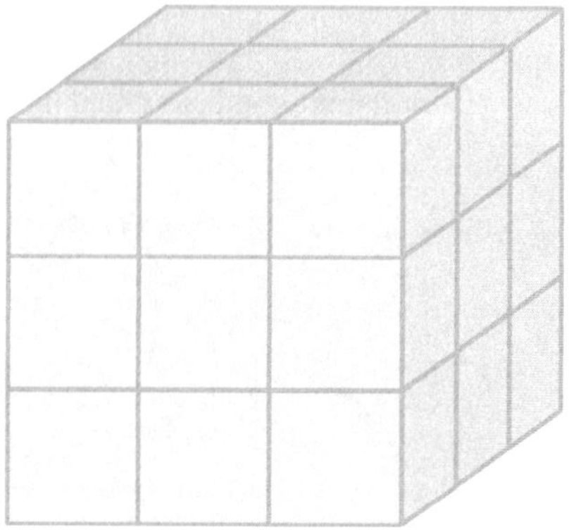

A naturalist measured the wingspan of blue jays and created the line plot below.

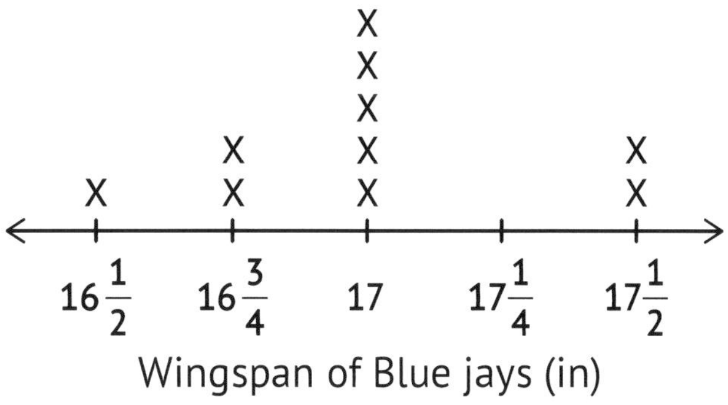

16. How many birds did he measure?

A. 8 **B.** 10 **C.** 12 **D.** 15

17. What is the difference between the largest wingspan and the smallest wingspan?

A. $\frac{1}{2}$ inch **B.** $1\frac{1}{4}$ inch **C.** 1 inch **D.** $\frac{3}{4}$ inch

18. Use the area model to determine the next partial quotient.

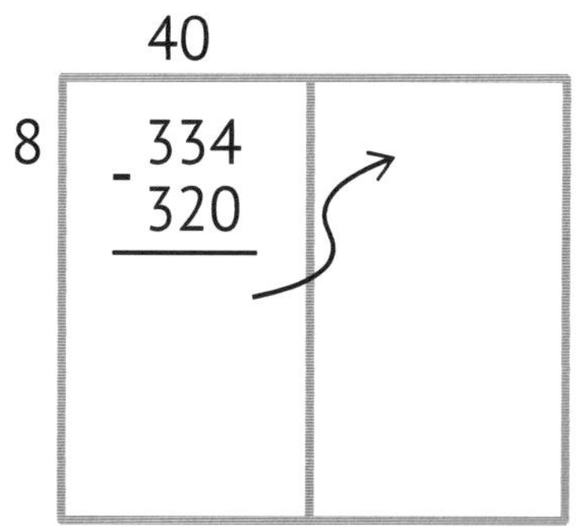

A. 16
B. 8
C. 1
D. 4

GO ON

19. Which expression would yield the smallest product?

A. $156 \times \frac{2}{3}$

B. $156 \times \frac{7}{5}$

C. $156 \times \frac{4}{4}$

D. $156 \times \frac{6}{4}$

20. Michael wants to store his dice. Each die is 1 cubic centimeter. He finds a box that is 5 cm long, 6 cm wide, and 4 cm tall. How many dice can he fit in the box?

A. 30 dice

B. 100 dice

C. 120 dice

D. 300 dice

21. Solve the expression.

$$5 + \left(\frac{34}{2}\right) \times 8 = \_\_\_\_\_\_$$

A. 114

B. 141

C. 176

D. 167

22. What is the value of the point on the number line? Which tenths place value does it round to?

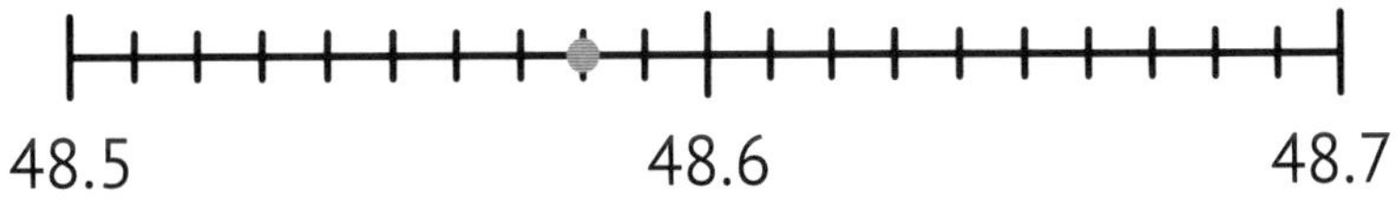

A. 48.51 rounds to 48.5

B. 48.57 rounds to 48.6

C. 48.58 rounds to 48.6

D. 48.59 rounds to 48.6

23. Angelo is painting his house. He uses $\frac{3}{5}$ a gallon of paint on Wednesday. He uses $\frac{3}{10}$ of a gallon on Thursday. How much paint did he use across both days?

A. $\frac{6}{15}$ gallon

B. $\frac{9}{10}$ gallon

C. 1 gallon

D. $\frac{6}{10}$ gallon

24. When Stuart went to the store he had \$72.37. When he left he had \$45.92. How much money did Stuart spend at the store?

A. \$26.45

B. \$25.45

C. \$24.65

D. \$25.55

GO ON

25. Jerry brought home leftover pizza. There was $\frac{6}{8}$ of the pizza left when he brought it home. He ate $\frac{1}{3}$ of the leftover pizza for dinner. How much of the total pizza did Jerry eat for dinner?

A. $\frac{1}{4}$ of the whole pizza

B. 5 slices

C. $\frac{7}{11}$ of the whole pizza

D. $\frac{8}{18}$ of the whole pizza

26. If 4 pies are divided into pieces that are each $\frac{1}{8}$ of a whole, how many pieces would there be?

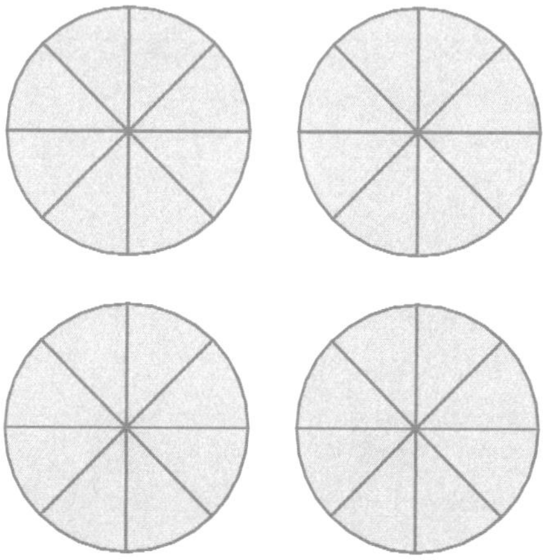

A. 2 pieces

B. 4 pieces

C. 16 pieces

D. 32 pieces

27. A teacher is creating cubbies for her classroom by nailing boxes together. Each box is 2 feet long, 2 feet wide, and 1 foot high. She is joining together 16 boxes. What is the total volume of storage space she is creating with the cubbies?

A. 64 cubic feet

B. 60 cubic feet

C. 20 cubic feet

D. 12 cubic feet

28. Owen was playing at an arcade and noticed that the longer a person played a game, the more tickets they received. He charted this pattern on a coordinate plane.

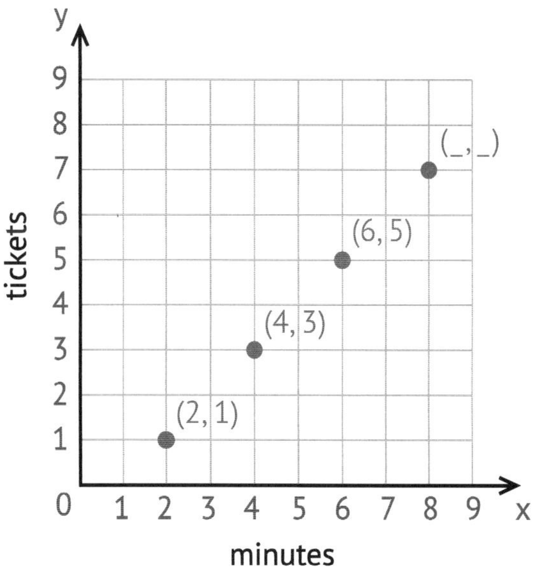

If a person plays the game for 8 minutes, how many tickets will they receive?

A. 6

B. 7

C. 8

D. 9

29. There are 6 friends at a pizza parlor. Each has $\frac{3}{4}$ the amount needed to buy a pizza. If they pool their money together, what is the greatest number of pizzas they can buy?

A. 2 pizzas

B. 3 pizzas

C. 4 pizzas

D. 5 pizzas

30. Use the number line to solve.

$$\frac{1}{10} \div 3 = \_\_\_\_\_\_$$

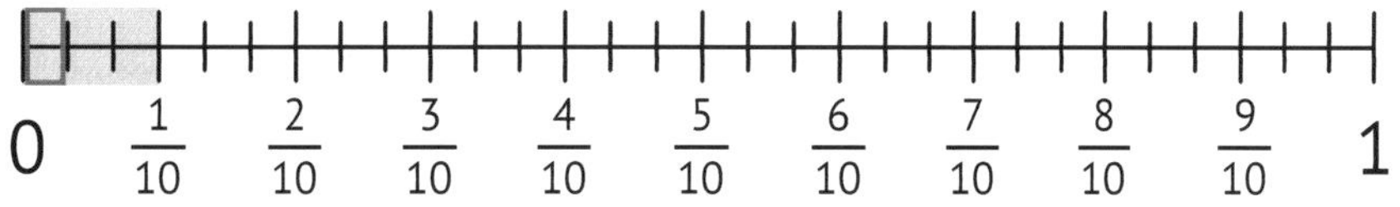

A. 30

B. $\frac{1}{30}$

C. $\frac{3}{30}$

D. 3

GO ON

Session 2

31. Susan is making curtains for her bedroom. One panel is $6\frac{2}{5}$ feet long and 3 feet wide. What is the area of the panel?

A. 64 feet

B. $19\frac{1}{5}$ feet

C. $12\frac{1}{2}$ feet

D. 60 feet

32. A hotel is getting a shipment of fresh linens for the beds. There are 24 bedsheets in one box and they received 56 boxes. How many sheets did they receive?

A. 1,424

B. 1,344

C. 1,324

D. 1,342

33. The following figure shows a unit fraction divided by a whole number. Write an expression that matches the figure below. Write a word problem that would match the expression and figure.

____

____

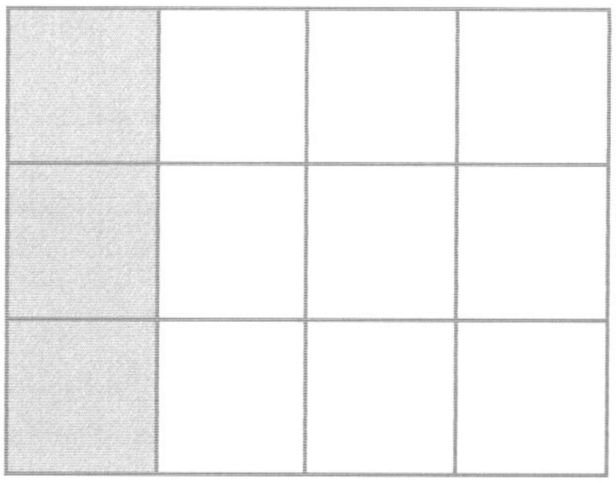

GO ON

34. Jerome had both strawberry and cookie cake at his birthday party. There was $\frac{3}{8}$ of strawberry cake leftover. There were $\frac{14}{24}$ of cookie cake leftover. How much cake did he have all together?

Use the visual model to show your work.

There was ________ leftover cake.

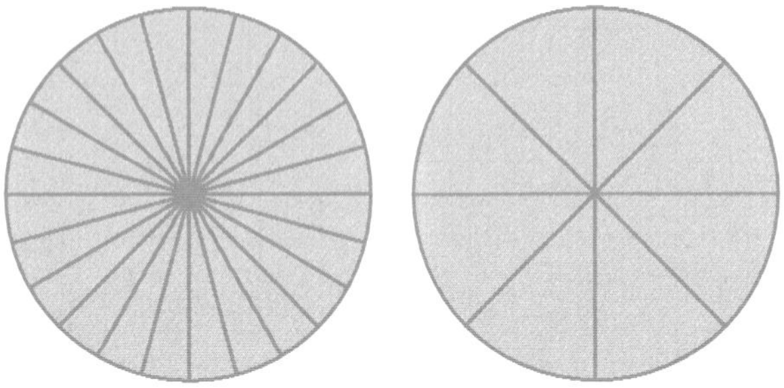

35. A class is having a party. They are drinking soda at their party. There are 12 fl oz of soda in one can. Each cup has the capacity of one actual cup. How many cups can they fill with 12 cans of soda?

1 cup = 8 fl oz

A. 144

B. 20

C. 64

D. 18

36. Use the area model below to solve the equation.

____

____

37. Grace makes candles for a local craft fair. She wants to test her candles to determine the approximate number of hours that her candles would burn. Create a line plot to show the time that each candle spent burning.

$$16\frac{3}{4}, 17\frac{1}{2}, 17\frac{1}{8}, 16\frac{3}{4}, 17, 16\frac{7}{8}, 16\frac{1}{2}, 17, 17\frac{1}{4}, 16\frac{1}{4}$$

38. Stephanie owns a garden shop. She is selling potted plants for a spring sale. She bought 832 ounces of potting soil. She has 40 flower pots to fill. Which equation will help her estimate how much potting soil she has for each flower pot?

A. 4 x 2 = 8

B. 40 x 20 = 80

C. 40 x 200 = 800

D. 400 x 200 = 8,000

39. Peter is preparing his garden for the spring. He adds $3\frac{5}{6}$ lbs of black potting soil and $\frac{3}{4}$ lbs of enriched potting soil into a planter. How much soil is in the planter?

Draw a visual model to represent the soil.

40. Which scenario would require you to determine the volume of a solid?

A. Covering a wall with wallpaper

B. Measuring the distance traveled per hour

C. Figuring out how much stuff can fit inside a moving box

D. Doubling a recipe

41. Thomas and Andy are painting a large wall together. Each of them is responsible for painting $\frac{1}{2}$ of the wall. Thomas has only finished painting $\frac{1}{4}$ of his $\frac{1}{2}$. How much of the wall has Thomas painted? Create a visual model to show your work.

42. Which expression would yield the least product?

A. $834 \times \frac{1}{2}$ **B.** $834 \times \frac{4}{3}$ **C.** $834 \times \frac{7}{7}$ **D.** $834 \times \frac{3}{2}$

43. Zoe is creating a mural of photographs for an art project. Each photo is $7\frac{1}{4}$ inches wide. Her mural is 18 photos wide. How wide is the mural?

A. 522 inches

B. $120\frac{1}{3}$ inches

C. $130\frac{1}{2}$ inches

D. 252 inches

44. Isaac is trying to solve the equation 32 x 1,000.

Solve the equation and explain how knowing the number of zeros in a multiple of ten can help to determine the product.

____

____

45. Which expression matches the phrase below?

The quotient of twelve and six times 15 minus 4

A. $\frac{12}{6} \times 15 - 4$

B. $(12 \times 6) \times 15 - 4$

C. $12 - 6 \times 15 - 4$

D. $\left(\frac{12}{6}\right) \times 15 - 4$

GO ON

Test 6: Session 1

Tips:

Be sure to read each question carefully and work through the problem step by step before choosing the correct answer.

Suggested Time:

Session 1: 55 to 65 Minutes
Session 2: 60 to 75 minutes

Students are **NOT** permitted to use a calculator.

1. Lester is passing out Valentines to his classmates. He got everyone a box of candy hearts. Each box has 125 hearts in it. He has 24 classmates. How many candy hearts did Lester give to his classmates?

A. 2,200
B. 4,500
C. 3,000
D. 450

2. Audrey has 5 children. Each of her children share some of their Halloween candy with her. They each give her $\frac{2}{6}$ of a candy bar. How much of a whole candy bar did Audrey get?

A. 10 pieces
B. $\frac{2}{18}$ of a candy bar
C. 2 candy bars
D. $1\frac{4}{6}$ candy bars

3. Greg ran 1 mile in 8 min 29 seconds. He said every second was excruciating. How many seconds did he run?

1 min = 60 seconds

A. 609 seconds
B. 489 seconds
C. 409 seconds
D. 509 seconds

GO ON

4. What is the value of the following expression?

$$\frac{21}{1{,}000} = \_\_\_\_\_\_$$

A. 2.1 **B.** .21 **C.** .021 **D.** .0021

5. Use the model to show the following equation.

$$\frac{8}{10} - \frac{2}{5} = \_\_\_\_\_\_$$

A. $\frac{2}{5}$ **B.** $\frac{2}{10}$ **C.** $\frac{4}{5}$ **D.** $\frac{4}{10}$

6. Which formula would help to determine the volume of a solid figure?

A. Length x width
B. Length x width x height
C. Length x width x 2
D. (Length x 2) + (width x 2) + (height x 2)

7. Use the area model to determine the next partial quotient.

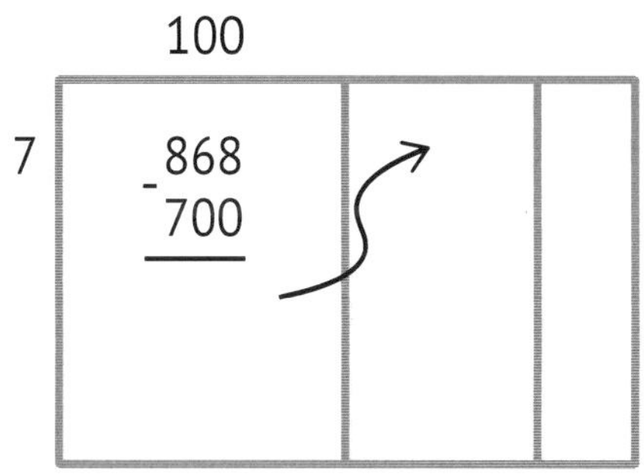

A. 24
B. 40
C. 20
D. 14

8. Biologists were studying microbes in a pond. One year they found 34 microbes in their water sample. The next year they found 3,400 microbes in their water sample. Which expression shows the change between the water samples?

A. 34×10^1
B. 34×10^2
C. 34×10^3
D. 34×10^4

9. Dorothy is laying out square blankets for a picnic. Each blanket is 4 square meters. The picnic area is 6 feet long by $5\frac{3}{4}$ feet wide. How many blankets would she need to cover the entire area?

A. 7 blankets
B. 8 blankets
C. 9 blankets
D. 10 blankets

10. Which is true of squares?

A. All squares have equal sides.
B. All squares are parallelograms.
C. All squares have right angles.
D. All of the above.

11. What is the value of 569.147 when rounded to the tenths place?

A. 569.1
B. 569.2
C. 570
D. 560

12. Ruby is mixing paint. She adds $\frac{2}{8}$ cups of orange, $\frac{1}{4}$ cups red, $\frac{3}{8}$ cups of white. Estimate the amount of paint that she mixed.

A. Less than $\frac{1}{2}$ cup
B. About a half cup
C. More than a half cup, less than a whole
D. More than a whole cup

13. Each square represents a cubic inch. Use the figure below to determine the volume of cubic inches.

A. 24 cubic inches

B. 46 cubic inches

C. 48 cubic inches

D. 50 cubic inches

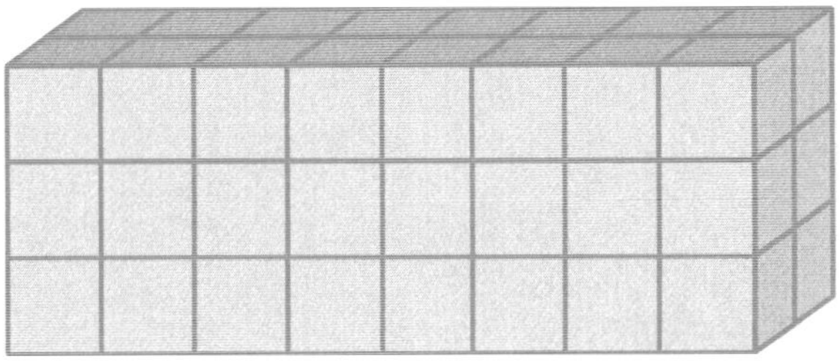

14. A truck can get 25 miles per gallon. The gas tank can hold 23 gallons when it is full. How many miles can the truck travel with a full tank of gas?

A. 575

B. 125

C. 555

D. 285

15. Which expression would create the greatest product?

A. $532 \times \frac{1}{2}$

C. $532 \times \frac{5}{5}$

B. $532 \times \frac{6}{5}$

D. $532 \times \frac{2}{9}$

GO ON

16. Which standard form number matches the value of the word form?

Three-thousand seven-hundred twenty-two and nine thousandths

A. 3,722.9
B. 3,722.09
C. 3,722.009
D. 3,722.0001

17. There were 7 mini pies at a buffet. Eight people wanted to try some. How much pie would each person get?

A. 15 pieces
B. 15 bites
C. $\frac{1}{56}$
D. $\frac{7}{8}$

18. Agnes is looking for boxes to store candles. She needs a box with a volume of at least 30 cubic inches. She finds a box that is 6 inches long, 4 inches wide and 3 inches tall. What is the volume of the box?

A. 24 in^3
B. 36 in^3
C. 64 in^3
D. 72 in^3

19. Solve using the visual model.

$$\frac{1}{5} \div 4 = \_\_\_\_\_\_$$

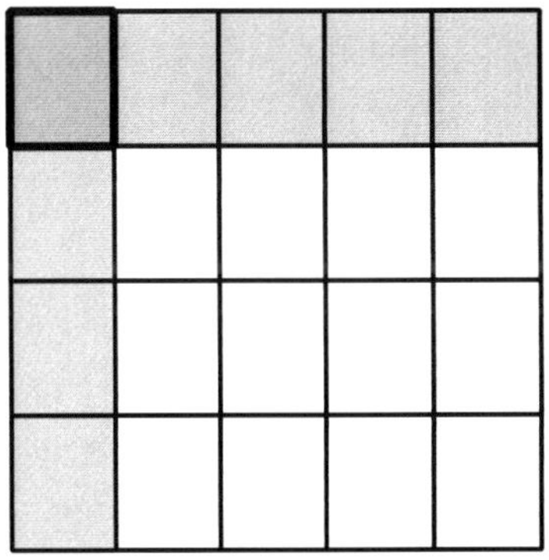

A. 20

B. $\frac{1}{20}$

C. 8

D. $\frac{8}{20}$

20. Anna was measuring the width of snowflakes.

Snowflake A	Snowflake B	Snowflake C	Snowflake D
.2	.1	.02	.05

Which of the snowflakes has the smallest width?

A. Snowflake A

B. Snowflake B

C. Snowflake C

D. Snowflake D

21. Solve.

$$\frac{4}{5} + \frac{2}{3} + \frac{7}{15} = \_\_\_\_\_\_$$

A. $1\frac{14}{15}$

B. $\frac{28}{15}$

C. $\frac{13}{23}$

D. $1\frac{12}{15}$

GO ON

22. Isaac is making baby food pouches. He made 5 cups of pureed banana and strawberries. Each pouch can hold 5 tablespoons. How many pouches can he fill?

1 cup = 16 tbsp

A. 80 pouches
B. 16 pouches
C. 20 pouches
D. 40 pouches

Matt owns a car lot and wants to plan where to place the cars. He measured the length of the cars and created a line plot.

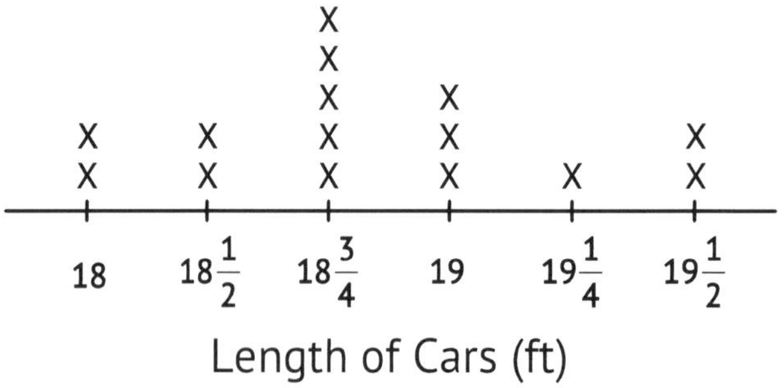

23. What is the difference between the longest car and the shortest car?

A. 2 feet
B. $1\frac{3}{4}$ feet
C. 1 foot
D. $1\frac{1}{2}$ feet

24. How many cars did he measure?

A. 10
B. 12
C. 15
D. 20

25. A politician got 32,100 votes. Their opponent got $\frac{1}{100}$ the number of votes. How many votes did their opponent get?

A. 3,210,000
B. 321,000
C. 3,210
D. 321

26. Eric bought tiles to put on his kitchen walls. He wants to put tiles on 3 of his walls. He bought 954 tiles. Which equation will help him estimate how many tiles will go on each wall?

A. 3 x 3 = 9
B. 3 x 30 = 90
C. 3 x 300 = 900
D. 3 x 3,000 = 9,000

27. Put the expressions in order from least to greatest.

$$724 \times \frac{3}{4} \qquad 724 \times \frac{9}{5} \qquad 724 \times \frac{1}{5} \qquad 724 \times \frac{3}{3}$$

A. $724 \times \frac{9}{5}$; $724 \times \frac{1}{5}$; $724 \times \frac{3}{4}$; $724 \times \frac{3}{3}$

B. $724 \times \frac{1}{5}$; $724 \times \frac{3}{4}$; $724 \times \frac{9}{5}$; $724 \times \frac{3}{3}$

C. $724 \times \frac{1}{5}$; $724 \times \frac{3}{4}$; $724 \times \frac{3}{3}$; $724 \times \frac{9}{5}$

D. $724 \times \frac{3}{3}$; $724 \times \frac{9}{5}$; $724 \times \frac{1}{5}$; $724 \times \frac{3}{4}$

GO ON

28. Kevin has 24 pairs of shoes. Mark has $\frac{3}{4}$ the number of pairs as Kevin. How many pairs of shoes does Mark have?

A. 12 pairs
B. 15 pairs
C. 18 pairs
D. 20 pairs

29. Which expression matches the phrase below?

Eight times the sum of two and six divided by four

A. $8 \times 2 + \frac{6}{4}$
B. $8 \times \frac{(2 + 6)}{4}$
C. $(8 \times 2) + \frac{6}{4}$
D. $8 \times 2 + \left(\frac{6}{4}\right)$

30. Classify the triangle based on its sides and angles.

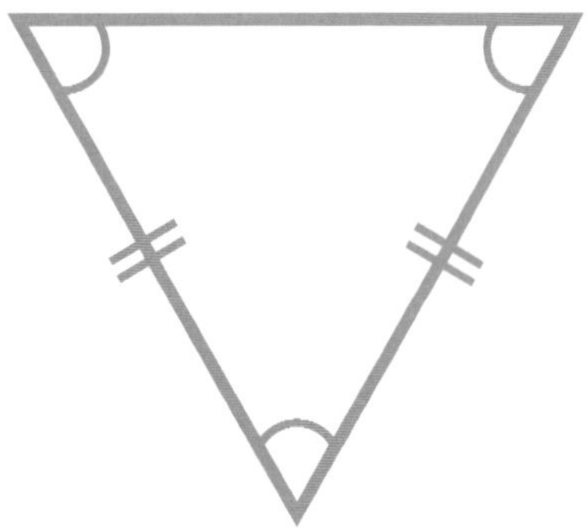

By Side	By Angle
Equilateral - has three equal sides	Acute - has three angles less than 90°
Isosceles - has two equal sides	Right - has one angle equal to 90°
Scalene - has no equal sides	Obtuse - has one angle greater than 90°

A. Equilateral acute
B. Isosceles acute
C. Scalene obtuse
D. Isosceles obtuse

Session 2

GO ON

31. Graham is on a long run. When he left the house, his water bottle was $\frac{3}{4}$ full. At the end of the run he'd consumed $\frac{1}{2}$ of the water in his bottle. How much of the whole bottle did he drink during his run?

A. $\frac{3}{8}$ bottle

B. $\frac{4}{6}$ bottle

C. $\frac{1}{2}$ bottle

D. $\frac{3}{4}$ bottle

32. Emerson is planning a trip across the country. He will need to travel 2,255 miles. He wants to make the trip in 5 days. He says that knowing the math fact 5 x 400 would help him estimate the number of miles he needs to travel in one day. How can knowing a multiplication equation help him divide? How many miles does he need to travel each day?

Show your work and explain.

____

____

33. Kelly wants to mail a birthday gift. The post office charges 2 cents per oz. The present she wants to send is 14 lbs. How much will she need to send to ship it?

A. $4.45

B. $4.50

C. $4.85

D. $4.48

34. A restaurant is getting a shipment of ingredients. They recieve 30 boxes of pasta. Each box has 24 ounces in it. Ramsey calculates that they will be receiving 72 ounces of pasta. What error did he make when multiplying? How many ounces of pasta will they actually get?

Show your work and explain.

____

____

35. Jeff had $5.37 and went to the store. He spent $2.49 cents. How much money did he have left? Use the visual model to show subtraction.

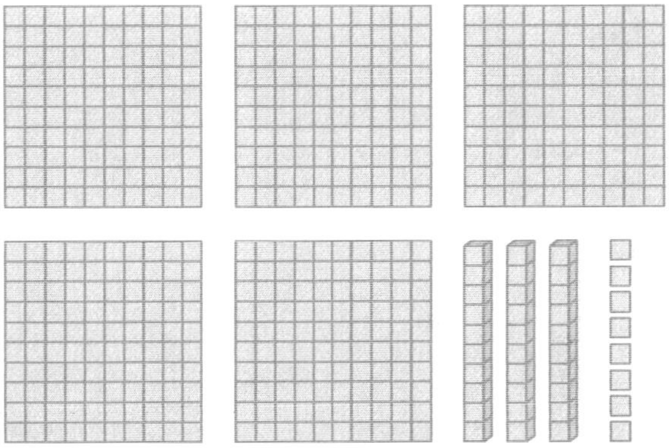

A. $2.99 **C.** $2.88

B. $3.99 **D.** $3.88

36. Solve.

$$8 \div \frac{1}{4} = \_\_\_\_\_$$

A. 2 **C.** 24

B. 32 **D.** 16

GO ON

37. Harry is in a stone skipping contest. The first time the stone skips it moves $3\frac{3}{4}$ feet before hitting the water again. The second time the stone skips it travels $\frac{3}{8}$ of a foot before it hits the water again. How far did it travel across both skips? Use the number line to represent the movement of the stone.

38. Use the shape below to label a figure with a volume of 24 cubic units.

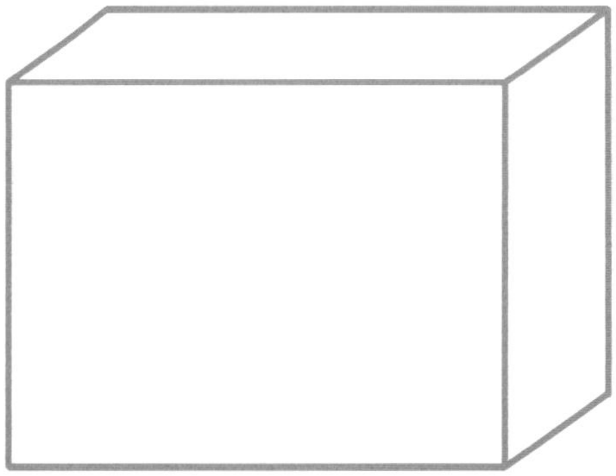

39. Look for a pattern in the data set below.

x	y	Coordinate Points
0	10	(0, 10)
1	8	(1, 8)
___	6	(___, 6)
3	___	(3, ___)
___	2	(___, 2)

What rule helped you determine the x coordinates? ____________________

____

What rule helped you determine the y coordinates? ____________________

____

40. What is the missing partial quotient?

$$
\begin{array}{r}
2 \\
\square \\
500 \\
6 \overline{)\ 3{,}492} \\
-\ 3{,}000 \\
\hline
492 \\
-\ 480 \\
\hline
12 \\
-\ 12 \\
\hline
0
\end{array}
$$

A. 800 **B.** 80 **C.** 8 **D.** 7

41. Ryan is riding his bike around town to run errands. All of his stops are along one road. He lives $2\frac{1}{2}$ miles from the grocery store. The grocery store is $\frac{7}{12}$ of a mile from the post office. The post office is $\frac{4}{6}$ of a mile from the bank. How far does Ryan have to bike to complete his errands and then get home?

A. 5 miles **B.** $5\frac{3}{5}$ miles **C.** 6 miles **D.** $7\frac{1}{2}$ miles

42. A prom dress store wants to get a red carpet for customers to try on their gowns. The red fabric is $12/yard. The store is 48 feet long. How many yards of red fabric should they buy?

A. 12 yards
B. 16 yards
C. 20 yards
D. 24 yards

GO ON

43. Compare the decimals by using <, >, =.

A. 82.15 ________ 82.105

B. 91.1 ________ 91.10

C. 45.1 ________ 4.51

D. 92.151 ________ 92.251

44. Angela says that 723 x $\frac{1}{4}$ would be greater than 723 because when you multiply numbers, the product is always larger than the factors. Oscar says that the product would be less than 723 because $\frac{1}{4}$ is less than 1 whole number. Who is correct? How do you know?

Explain your thinking.

____

____

45. Raj likes to bake, but he is running out of ingredients. He is only making $\frac{1}{4}$ of the recipe. The recipe calls for $2\frac{1}{4}$ cups of milk. How much does he need to make $\frac{1}{4}$ of the recipe?

A. $1\frac{1}{2}$ cups

B. $\frac{9}{16}$ cups

C. $1\frac{3}{4}$ cups

D. $\frac{3}{4}$ cups

Test 7: Session 1

Tips: Be sure to read each question carefully and work through the problem step by step before choosing the correct answer.

Suggested Time:

Session 1: 55 to 65 Minutes
Session 2: 60 to 75 minutes

Students are **NOT** permitted to use a calculator.

1. Steven is trying to watch a movie, but he keeps getting interrupted. He watched $\frac{1}{8}$ of the movie on Thursday. He watched $\frac{2}{5}$ of the movie on Friday. About how much of the movie has he watched so far?

A. About a quarter
B. About a half
C. About three quarters
D. Close to a whole

2. Meagan works at a bakery for dogs. She is selling dog treats for $4 per pound. Each treat is 2 ounces. She made 36 treats. How many pounds of dog treats can she sell?

A. 4 lbs
B. $4\frac{1}{2}$ lbs
C. 5 lbs
D. $5\frac{1}{2}$ lbs

3. Put the numbers in order from greatest to least.

87.101 **87.10** **87.01** **87.11**

A. 87.11; 87.101; 87.10; 87.01
B. 87.101; 87.10; 87.01; 87.11
C. 87.01; 87.10; 87.101; 87.11
D. 87.11; 87.01; 87.101; 87.10

4. There were 6 monkeys at the zoo. The zookeeper only had 5 bananas. They wanted to make sure that each monkey got an equal amount. How much will each monkey get?

A. 11 pieces

B. $\frac{6}{5}$ of a banana

C. $\frac{5}{6}$ of a banana

D. $\frac{1}{11}$ of a banana

5. Solve the expression below.

$$45 + \frac{(3 \times 8)}{6} = \_\_\_\_\_\_$$

A. 50 **B.** 48 **C.** 49 **D.** 64

6. Stephen owns a car wash. He gets \$3.25 every time someone buys a carwash. How much money does he make if he has 9 customers in one day?

Draw base ten blocks to multiply.

A. \$30.25

B. \$29.25

C. \$31.25

D. \$28.25

7. Which expression matches the figure below?

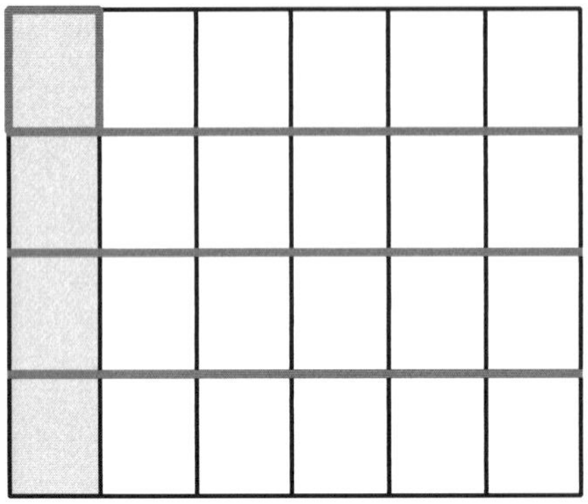

A. $\frac{1}{6} \div 4$ **B.** $\frac{1}{4} \div 6$ **C.** $\frac{1}{4} \div 4$ **D.** $\frac{1}{6} \div 6$

8. What is the quotient of the figure above?

A. $\frac{1}{4}$ **B.** $\frac{1}{6}$ **C.** $\frac{1}{24}$ **D.** $\frac{1}{10}$

9. Carl is reading a book series. During the first week, he read $\frac{6}{7}$ of a book. During the second week, he read $\frac{1}{2}$ of a book. During the third week, he read $\frac{3}{5}$ of a book. About how much of the series has he read?

A. About one book
B. About one and a half books
C. About two books
D. About two and a half books

10. Sarah has 842 pictures and 4 photo albums. She wants to divide the pictures evenly between the 4 albums.

Which multiplication equation would help her determine how many pictures should go in each album?

A. 4 x 2 = 8

B. 4 x 20 = 80

C. 4 x 200 = 800

D. 4 x 2,000= 8,000

11. How many photos would go in each album? Use the area model below to divide.

A. 210

B. 210 R 2

C. 212

D. 210 R 4

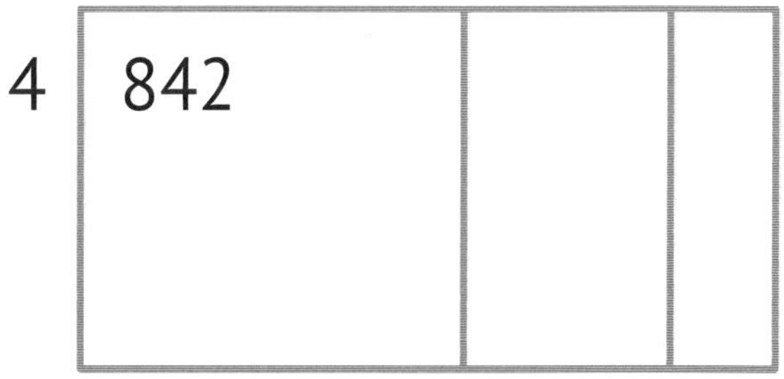

12. Tony wants to fill a box with cereal. He is trying to determine how much cereal could fit in the box. The pieces of cereal are about 1 cm cubed. The box is 12 cm long, 3 cm deep, and 18 cm tall. What is the volume of the cereal box?

A. 650 cm^3

B. 684 cm^3

C. 648 cm^3

D. 560 cm^3

GO ON

13. Abby earned money by doing some chores. She drew base-ten blocks to represent the money she had and the money she earned. Which expression matches her drawing?

A. \$5.40 + \$3.54

B. \$5.36 + \$3.54

C. \$5.37 + \$3.54

D. \$5.37 + \$3.63

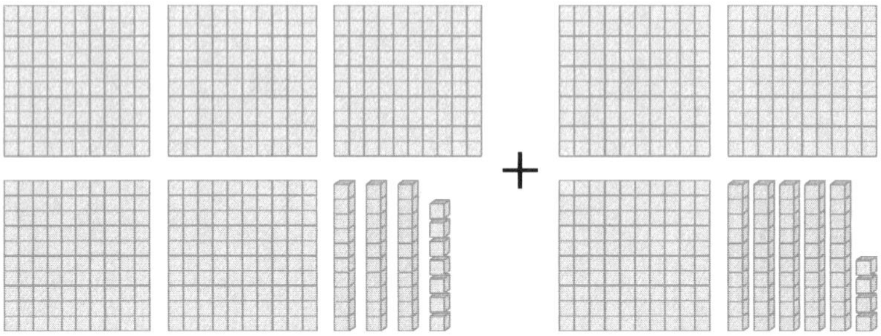

14. How much money does Abby have after doing chores?

A. \$8.92

B. \$8.91

C. \$8.90

D. \$8.88

15. Determine the volume of cubic units.

A. 150

B. 130

C. 120

D. 90

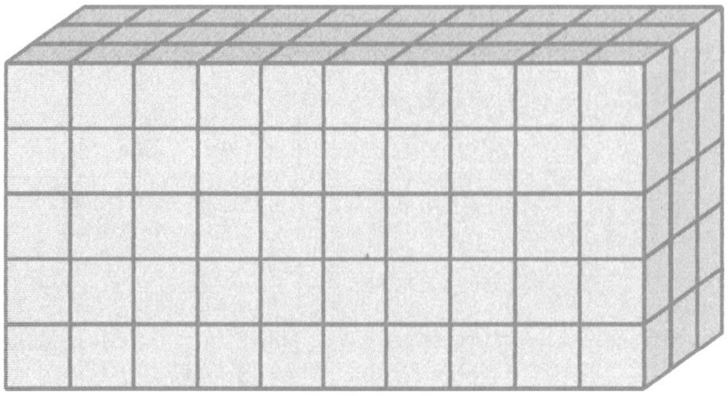

16. Which attribute is true of all triangles?

A. All triangles have at least one acute angle

B. All triangles have equal sides

C. All triangles have either a right or obtuse angle

D. All triangles have parallel sides

A farmer is examining the growth of young apple trees in his orchard. He creates a line plot to organize his data.

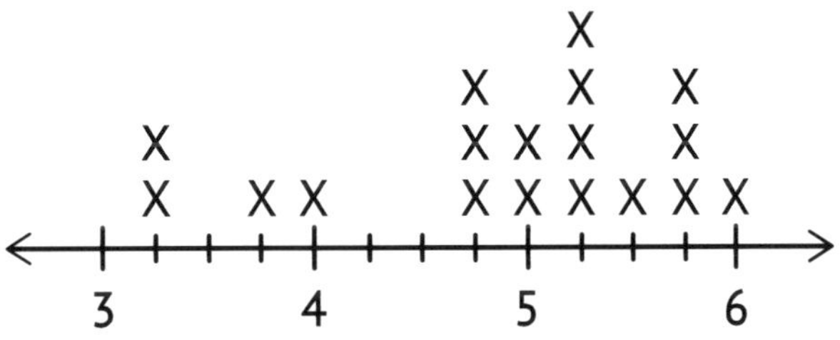

Diameter of tree trunks (in)

17. How many trees did he measure?

A. 15 trees

B. 10 trees

C. 18 trees

D. 20 trees

18. What is the difference between the diameter of the largest tree and the smallest tree?

A. 3 inches

B. $3\frac{1}{3}$ inches

C. 2 inches

D. $2\frac{3}{4}$ inches

GO ON

19. Which expression would yield the number 73,000?

A. 73×10^2
B. 73×10^3
C. 73×10^4
D. 73×10^5

20. Which is not a unit of measurement that would be used to determine the volume of a solid?

A. Degrees
B. Inches
C. Feet
D. Centimeters

21. Which number would round to 46.2?

A. 46.34
B. 46.051
C. 46.15
D. 46.09

22. Which expression matches the phrase below?

Sixteen plus twenty-four divided by the product of two and four

A. $\frac{(16 + 24)}{2} \times 4$

B. $16 + \frac{24}{2} \times 4$

C. $16 + \frac{24}{(2 \times 4)}$

D. $\frac{(16 + 24)}{(2 \times 4)}$

23. George is cutting wood in his woodshop. The first piece of wood he cuts is 40 inches wide. The second piece of wood is .4 inches wide. How much larger was the first piece?

A. 10 times larger
B. 100 times larger
C. 1,000 times larger
D. 10,000 times larger

24. Maggie is working on a project. On Monday she finished $\frac{1}{4}$ of the project. On Tuesday she completed $\frac{5}{12}$ of the project. How much does she still need to complete?

A. $\frac{1}{2}$

B. $\frac{1}{3}$

C. $\frac{1}{4}$

D. $\frac{1}{5}$

GO ON

25. Blake is getting ready to use his sailboat, but he needs to buy more rope. Rope is sold by the foot. He knows he needs 15 yards of rope. How many feet of rope does he need?

A. 12 feet

B. 30 feet

C. 45 feet

D. 60 feet

26. A garden store sold sunflower seed packets. They sold a total of 65 packets. Each packet has 205 seeds in it. How many sunflower seeds were sold by the store?

A. 13,300

B. 13,425

C. 13,325

D. 13,400

27. Rachel is shipping chocolates to a friend. The post office charges per ounce. She bought 12 chocolates. Each of them is $\frac{2}{5}$ of an ounce. How many ounces of chocolates does she need to ship?

A. 5 ounces

B. $4\frac{1}{2}$ ounces

C. $4\frac{4}{5}$ ounces

D. $4\frac{1}{4}$ ounces

28. Use the model to solve $\frac{2}{3} \times \frac{4}{5}$.

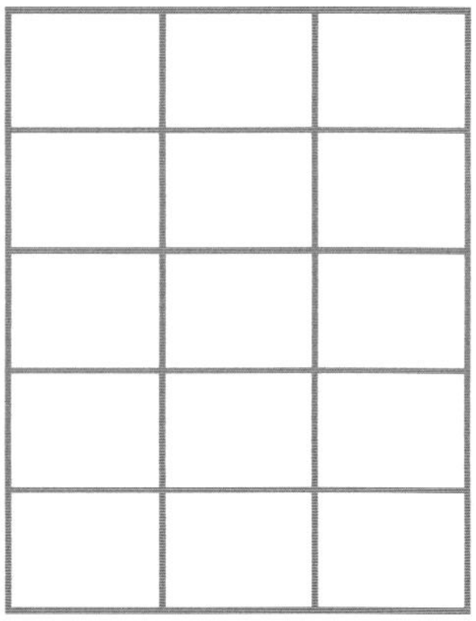

A. 15

B. $\frac{1}{15}$

C. $\frac{6}{15}$

D. $\frac{8}{15}$

29. Which equation would yield the greatest product?

A. $54 \times \frac{1}{3}$

B. $54 \times \frac{2}{5}$

C. $54 \times \frac{9}{9}$

D. $54 \times \frac{3}{2}$

30. Cecilia is making a costume. She bought a piece of fabric that is $2\frac{1}{3}$ yards by $5\frac{4}{5}$ yards. What is the area of the fabric?

A. $12\frac{8}{15}$

B. $13\frac{8}{15}$

C. $12\frac{9}{15}$

D. $13\frac{9}{15}$

GO ON

Session 2

31. A group of art students are making collages using different materials. There are 8 strips of fabric left. 9 students want to use fabric. How much fabric would they each get if they split it evenly?

A. 17

B. $\frac{1}{17}$

C. $\frac{8}{9}$

D. $\frac{9}{8}$

32. Use the number line to round 32.7 to the nearest whole number.

33. There is $\frac{1}{5}$ a pan of brownies leftover. Four friends decided to share it. How much of a pan would each friend get?

A. $\frac{1}{9}$

B. $\frac{4}{5}$

C. $\frac{1}{20}$

D. $\frac{1}{16}$

GO ON

34. A company is trying to gain more followers on their social media accounts. In 2019 they got 230 followers. In 2020 they have 2,300,000 followers. Compared to 2019, how many more followers did they have in 2020?

A. $\frac{1}{100}$ as many

B. $\frac{1}{1000}$ as many

C. 1,000 times as many

D. 10,000 times as many

35. A group of friends want to go on a vacation together. They found a trip that would cost $4,680. There are 9 friends in the group. How much would it cost per person? Use the area model to solve.

9	4,680		

36. $317 \times \frac{1}{4}$

Angela says that she knows before solving that the product would be less than 317. Is she correct? Explain how you know.

37. A grocery store supplier is receiving a shipment of bananas. Each box has 12 bunches in it. There are 45 boxes. How many bunches of bananas did the grocery store supplier receive?

A. 536

B. 600

C. 540

D. 560

38. Use the symbols <, >, = to compare the decimals.

A. 62.1 ________ 62.10

B. 54.32 ________ 54.302

C. 98.05 ________ 98.005

D. 472.13 ________ 473.12

39. Maria has $\frac{4}{5}$ of a leftover casserole. She eats $\frac{1}{3}$ of it for dinner. How much of the whole casserole does she have for dinner?

Use the visual model to solve.

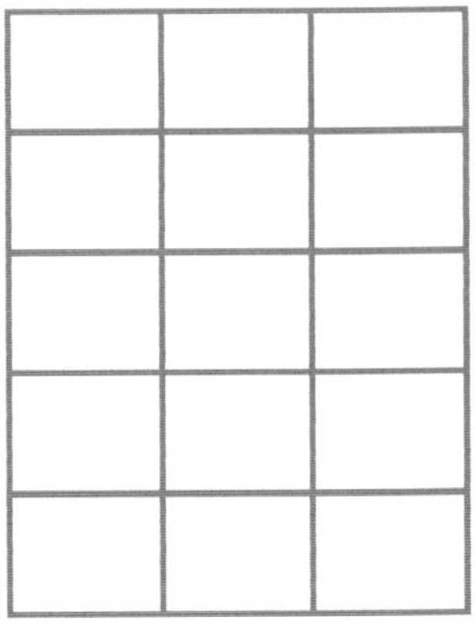

____

____

GO ON ▶

40. Solve the expression:

$$7 \times \frac{(4+5)}{(5-2)} = \_\_\_\_\_\_$$

A. 20

B. 21

C. 60

D. 61

41. Write a scenario to match the equation below and solve.

$$9 \times 36 = \_\_\_\_\_\_$$

____

____

42. A shelf on a bookshelf says it can support 5 lbs. Brian has a lot of trade novels. Each one weighs about 4 ounces. How many trade novels could the shelf support?

A. 50

B. 20

C. 70

D. 80

43. Toby made a goal to practice guitar for an hour during the week. On Monday he practiced $\frac{1}{6}$ of an hour. On Wednesday he practiced $\frac{1}{2}$ of an hour. How much more time does he need to practice to meet his goal?

A. $\frac{1}{3}$ hour

B. $\frac{1}{2}$ hour

C. $\frac{2}{3}$ hour

D. $\frac{1}{5}$ hour

44. A Lego manufacturer is boxing up Legos. There are 7,616 blocks that need to be packed into 8 boxes. How many blocks would go in each box? Use the partial quotient method to solve.

$$8 \overline{)\ 7{,}616}$$

A. 925

B. 952

C. 529

D. 592

45. Which phrase would match the expression below?

$$8 \times \frac{(7 - 2)}{4} = \_\_\_\_\_\_$$

A. Eight times seven minus two divided by four

B. Eight times five divided by four

C. Eight times the difference of seven minus two divided by four

D. Eight times seven minus two over four

GO ON

Test 8: Session 1

Tips: Be sure to read each question carefully and work through the problem step by step before choosing the correct answer.

Suggested Time:

Session 1: 55 to 65 Minutes
Session 2: 60 to 75 minutes

Students are **NOT** permitted to use a calculator.

1. On Tuesday it snowed 5.75 inches, but some of it melted overnight. On Wednesday there was 3.65 inches left. How much melted overnight? Use the model to subtract.

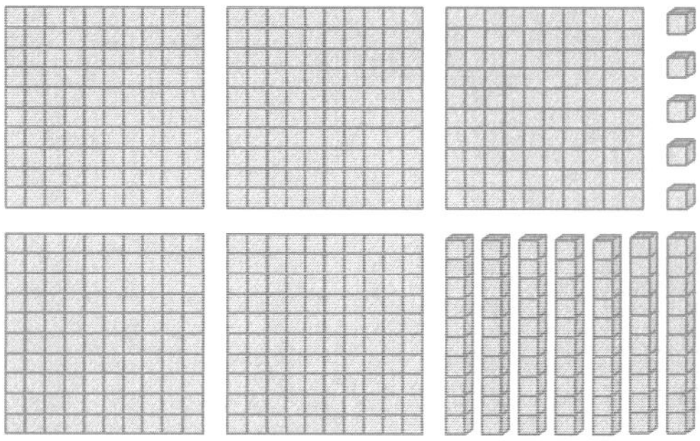

A. 3.1
B. 2.01
C. 2.1
D. 3.01

2. Put the expressions in order from greatest to least.

$$561 \times \frac{1}{4} \qquad 561 \times \frac{8}{8} \qquad 561 \times \frac{7}{5} \qquad 561 \times \frac{2}{5}$$

A. $561 \times \frac{7}{5}$; $561 \times \frac{8}{8}$; $561 \times \frac{2}{5}$; $561 \times \frac{1}{4}$

B. $561 \times \frac{8}{8}$; $561 \times \frac{2}{5}$; $561 \times \frac{7}{5}$; $561 \times \frac{1}{4}$

C. $561 \times \frac{2}{5}$; $561 \times \frac{7}{5}$; $561 \times \frac{1}{4}$; $561 \times \frac{8}{8}$

D. $561 \times \frac{8}{8}$; $561 \times \frac{1}{4}$; $561 \times \frac{7}{5}$; $561 \times \frac{2}{5}$;

3. What is the value of the decimal on the number line? What whole number would it round to?

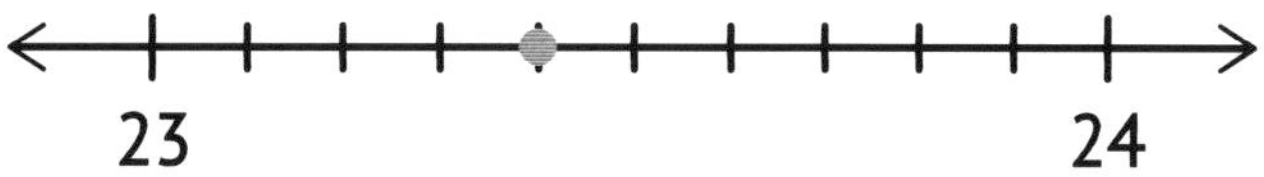

A. 23.5 would round to 24
B. 23.4 would round to 23
C. 23.05 would round to 24
D. 23.04 would round to 23

GO ON

An archeologist worked to collect 514 rock samples. They need to send the samples back to their lab. They can fit 12 samples in one box.

4. Which multiplication equation would help them determine how many boxes they would need?

A. 12 x 4 = 48
B. 12 x 40 = 480
C. 12 x 5 = 60
D. 12 x 50 = 600

5. How many boxes would the archeologist need? Use the area model below to solve.

A. 40 boxes
B. 41 boxes
C. 42 boxes
D. 43 boxes

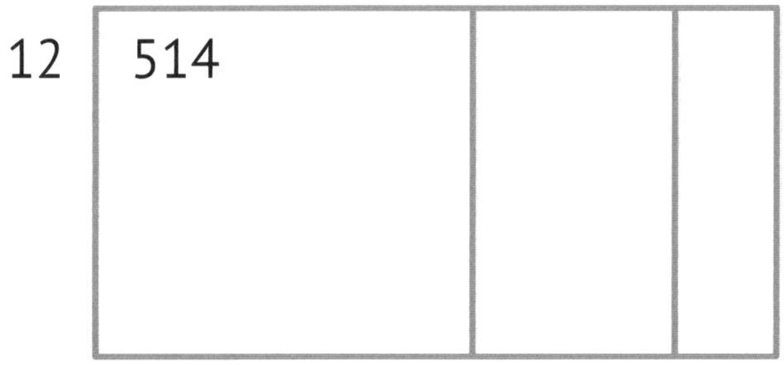

6. Art students are using construction paper to create artwork. There are 5 pieces of red construction paper left. There are 7 students that decide to split it evenly. How much red construction would each student get?

A. $\frac{7}{5}$ of a page
B. $\frac{5}{7}$ of a page
C. $\frac{1}{12}$ of a page
D. $\frac{1}{35}$ of a page

An engineer is creating a battery and testing the lifespan of its use. She created a line plot to organize her data.

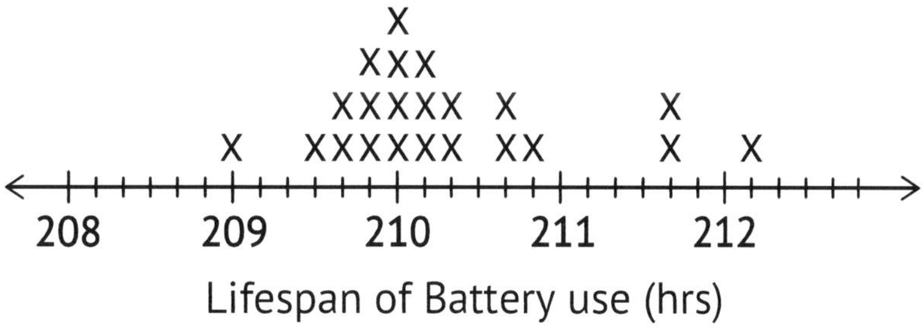

7. What is the mode of this data set?

A. 208 hrs

B. 209 hrs

C. 210 hrs

D. 211 hrs

8. How long did the battery with the longest lifespan last?

A. 210 hrs

B. 212 hrs

C. $211\frac{4}{6}$ hrs

D. $212\frac{1}{6}$ hrs

9. A book printer created 85 copies of a book. Each book had 215 pages. How many pages did they print?

A. 18,275

B. 18,215

C. 18,365

D. 18,285

GO ON

10. Mike is preparing for a snowstorm. He buys two gallons of milk. On Monday he uses $\frac{1}{4}$ of a gallon for cereal. On Tuesday he uses $\frac{1}{6}$ of a gallon when baking. On Wednesday he uses $\frac{1}{8}$ of a gallon to make hot cocoa. About how much milk did he use so far?

A. Less than a half gallon

B. About a half gallon

C. About one gallon

D. About one and a half gallons

11. Which expression matches the model below?

A. $\frac{4}{5} \times \frac{3}{4}$

B. $\frac{3}{5} \times \frac{3}{4}$

C. $\frac{3}{6} \times \frac{3}{4}$

D. $\frac{2}{5} \times \frac{3}{4}$

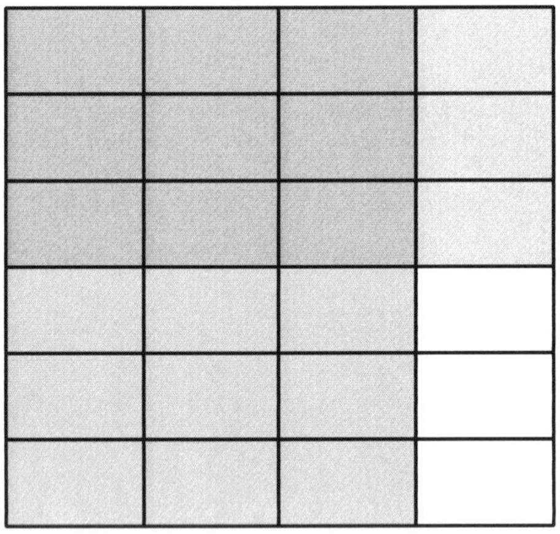

12. What is the product of the model above?

A. $\frac{9}{24}$

B. $\frac{1}{24}$

C. $\frac{3}{24}$

D. $\frac{6}{24}$

13. What is NOT true of every trapezoid?

A. Every trapezoid is a parallelogram
B. Every trapezoid is a quadrilateral
C. Every trapezoid has two obtuse and two acute angles
D. Every trapezoid has equal sides

14. What is the standard form of the decimal below?

Three-hundred eighty-three and eighty-three thousandths

A. 383,083
B. 383.083
C. 383.83
D. 383.0083

15. Susan has $\frac{3}{4}$ liters of cranberry juice. She needs to use $\frac{1}{2}$ of it for a recipe. How much cranberry juice is needed for the recipe?

A. $\frac{4}{6}$ liter
B. $\frac{3}{8}$ liter
C. $\frac{1}{2}$ liter
D. $\frac{2}{6}$ liter

GO ON ▶

16. Which does NOT describe the shape below?

A. It is a polygon
B. It is a quadrilateral
C. It is a parallelogram
D. It is a hexagon

17. Trey is stacking boxes. Each box is 4.25 inches tall. The cabinet is 30 inches tall. If he stacks 7 boxes, how much room will be left?

A. .5 inch
B. 1 inch
C. 1.25 inches
D. .25 inch

18. Use the model to solve.

$$\frac{1}{4} + \frac{5}{8} = \_\_\_\_\_\_$$

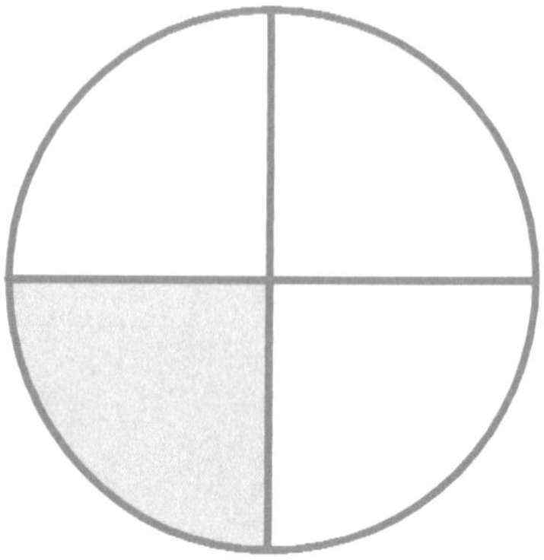

A. $\frac{6}{11}$

B. $\frac{6}{8}$

C. $\frac{7}{8}$

D. $\frac{3}{4}$

19. Count the cubic units to determine the volume.

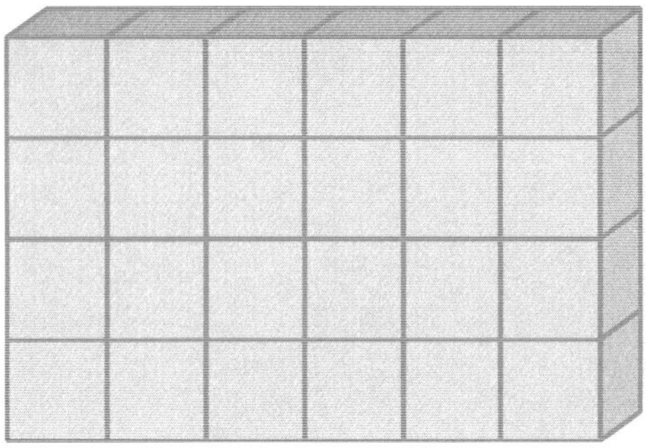

A. 34 cubic units
B. 30 cubic units
C. 24 cubic units
D. 28 cubic units

20. A movie producer wants to host an event for the premiere of their movie. They want to invite 725 people. There are 5 theaters available. How many people would be in each theater?

A. 150
B. 155
C. 140
D. 145

21. A restaurant is serving a chicken pot pie special. They bake 12 chicken pot pies. They divide them into eighths. How many pieces of pot pie can they serve?

A. 20 pieces
B. 80 pieces
C. 72 pieces
D. 96 pieces

GO ON

22. Compare the expressions:

$$4.25 \times 10^4 \ \_\_\_\_\_\_ \ 42.5 \times 10^3$$

A. <

B. >

C. =

D. +

23. Fred bought 1 lb of flour to use for baking projects. He used $\frac{1}{4}$ to make cookies. He used $\frac{2}{6}$ to make cake. He used $\frac{1}{8}$ lb to make pancakes. How much flour has he used altogether?

A. $\frac{17}{24}$ lb

B. $\frac{23}{24}$ lb

C. $\frac{21}{24}$ lb

D. $\frac{18}{24}$ lb

24. Jennifer is creating emergency kits for natural disasters. She wants to know the capacity of the box she's using. The box is 2 feet long, 2 feet wide, and 1 foot tall. How much space does she have if she has 4 boxes?

A. 4 ft^3

B. 8 ft^3

C. 16 ft^3

D. 20 ft^3

25. One candidate raised $700,000 for their campaign. Another candidate raised $700 for their campaign. How much more money did the first candidate raise?

A. They raised 10 times more.

B. They raised 100 times more.

C. They raised 1,000 times more.

D. They raised 10,000 times more.

26. Felicia is paying bills. She uses $\frac{2}{5}$ of her paycheck for rent. She uses $\frac{1}{3}$ of her paycheck for groceries. How much of her paycheck does she have left?

A. $\frac{1}{2}$

B. $\frac{1}{3}$

C. $\frac{11}{15}$

D. $\frac{4}{15}$

27. Someone used up almost all of the paper towels at a family gathering. There is only $\frac{1}{4}$ of a paper towel left. There are 5 people in the family. If they tear up the paper towel into equal pieces, how much of a whole paper towel will each family member get?

A. $\frac{1}{9}$

B. $\frac{1}{20}$

C. $\frac{1}{5}$

D. 1

GO ON ▶

28. Tommy has 2 tablespoons of sugar in his coffee. How many cups of sugar would he consume in 30 days?

1 cup = 16 tablespoons

A. 4 cups

B. $3\frac{3}{4}$ cups

C. $3\frac{1}{2}$ cups

D. 3 cups

29. What is the volume of a solid?

A. The measurement of the outside
B. The capacity of the inside
C. The weight
D. The number of sides

30. A lumber company needs to transport 88,000 lbs of lumber. How many tons of lumber do they need to transport?

1 ton = 2,000 lbs

A. 44 tons
B. 4,400 tons
C. 44,000 tons
D. 176,000 tons

Session 2

31. Shade the figure to show $\frac{1}{6} \times \frac{2}{3}$. What is the product?

____

____

32. A gardener is recording the height of tomato plants in his garden.

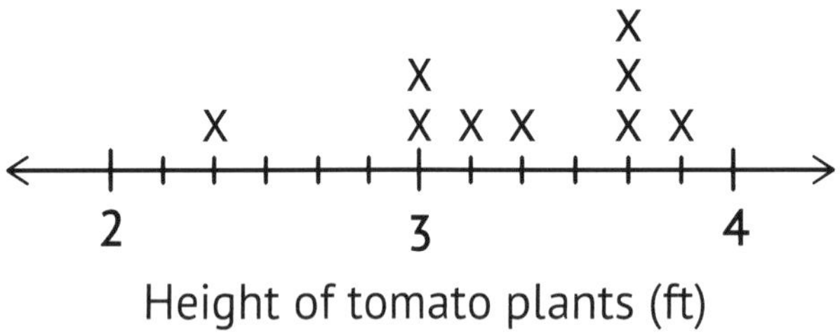

What is the difference between the tallest and shortest tomato plant?

A. $\frac{4}{5}$ foot **B.** $1\frac{1}{2}$ feet **C.** $\frac{3}{4}$ foot **D.** $1\frac{1}{4}$ feet

33. An engineer is designing a rollercoaster. She needs to make sure the tracks will support the cars. Each car weighs 214 lbs. There are 5 cars. How much weight should the track support?

A. 1,700 lbs **C.** 7,100 lbs

B. 1,070 lbs **D.** 7,010 lbs

34. Write a scenario to match the equation below and solve.

91 ÷ 7 = ________

____

____

35. David orders a pizza for dinner. That evening he eats $\frac{1}{3}$ of the pizza. The next day he eats $\frac{2}{8}$ of the pizza. About how much pizza did he eat?

A. A quarter
B. Less than half
C. About half
D. About a whole

36. A restaurant buys 6 gallons of broth to make soup. Each batch of soup needs $\frac{1}{5}$ of a gallon. How many batches of soup can be created from the 6 gallons of broth? Draw a model to show your thinking.

37. Oliver is trying to solve a multiplication problem, but he makes an error. Identify the error and explain how he made a mistake. Solve to get the correct answer.

$$\begin{array}{r} {}^{1} \\ 45 \\ \times\ 12 \\ \hline {}_{1}80 \\ +\ 450 \\ \hline 530 \end{array}$$

____

____

38. Laura is $4\frac{2}{3}$ feet tall. Amy is $\frac{1}{3}$ her sister's height. How tall is Amy?

A. $1\frac{5}{9}$ feet

B. $1\frac{2}{3}$ feet

C. $2\frac{7}{9}$ feet

D. $2\frac{1}{5}$ feet

39. James is trying to determine the volume of cubic inches that can fit in his box. He uses green cubes to represent the cubic inches.

He says that the box can fit 9 cubic inches. Is his conclusion correct? Why or why not?

____

____

40. A family is at the fair. The mom brought $9.75 for her three kids to spend on games and rides. How much money would each kid get if they split it evenly?

Draw base ten blocks to divide.

41. A teacher is giving out school supplies at the beginning of the year. She has 216 erasers to hand out. She has 18 students in her classroom. How many erasers will each student get? Use the area model to solve.

18	216	

____

____

42. Evan is evaluating the relationship between two sets of data.

X axis	Y Axis
0	0
1	5
2	10
3	15

What rule describes the relationship between the two sets?

A. Add one
B. Multiply by three
C. Add ten
D. Multiply by 5

GO ON

43. Jared is trying to solve 514.26 x 100. He tries to solve using an algorithm, but gets an incorrect answer.

$$\begin{array}{r} 514.26 \\ \times \quad 100 \\ \hline 00000 \\ 00000 \\ +\ 51426 \\ \hline 514.26 \end{array}$$

Is there an easier way to solve this problem? What is the correct solution? Explain how to use patterns to solve an equation.

____

____

44. Use the number line below to round 54.13 to the nearest tenths place.

45. A baby is learning to crawl. She crawls 18 feet. How many yards does she crawl?

A. 1 yard 6 feet
B. 6 yards
C. 4 yard 3 feet
D. 2 yards

ANSWER KEY

Question	Type	Key	Points	Standard	Cluster	Explanation
Session 1						
1	MC	D	1	CCSS.MATH. CONTENT. 5.NF.B.4	Numbers and Operations in Base Ten	Five groups of $\frac{1}{3}$ will yield $\frac{5}{3}$. That would simplify to $1\frac{2}{3}$ cups.
2	MC	A	1	CCSS.MATH. CONTENT. 5.NBT.A.3.A	Numbers and Operations in Base Ten	The parenthesis show the place values. Students can either recognize the pattern or solve and add.
3	MC	A	1	CCSS.MATH. CONTENT. 5.NF.B.6	Numbers and Operations Fractions	To solve, multiply the whole number by the fraction. Change the denominator of the fractions to subtract.
4	MC	A	1	CCSS.MATH. CONTENT. 5.NF.B.3	Numbers and Operations Fractions	Students must recognize that fractions can be represented with division.
5	MC	B	1	CCSS.MATH. CONTENT. 5.NBT.B.6	Numbers and Operations in Base Ten	To find the first partial quotient, students must recognize that the largest group of 12 they can make with 4100 is 300 because 3 time 12 is 36. (They must realize that 400 would be too large because 4 time 12 is 48.)
6	MC	C	1	CCSS.MATH. CONTENT. 5.NF.B.7	Numbers and Operations-Fractions	Students must recognize that the numerator of the answer is the shaded portion of the figure. They must count all sections to determine the denominator.
7	MC	D	1	CCSS.MATH. CONTENT. 5.NF.A.2	Numbers and Operations-Fractions	To solve, change the denominator before subtracting.
8	MC	D	1	CCSS.MATH. CONTENT. 5.NBT.B.5	Numbers and Operations in Base Ten	Students can solve using models, place value strategies, or the traditional algorithm.
9	MC	C	1	CCSS.MATH. CONTENT. 5.MD.B.2	Measurement and Data	To solve, change the denominators of the fractions to add all of the data points.
10	MC	A	1	CCSS.MATH. CONTENT. 5.MD.B.2	Measurement and Data	To solve, change the denominator of the fractions to subtract.
11	MC	C	1	CCSS.MATH. CONTENT. 5.MD.A.1	Measurement and Data	There are 108 inches of fabric in 3 yards. 108 divided by 3 is 36.
12	MC	B	1	CCSS.MATH. CONTENT. 5.NBT.A.4	Numbers and Operations in Base Ten	Two is in the tenths place. Since 8 is in the hundredths place, it indicates that the value is close to 3 tenths than it is to 2 tenths.

Question	Type	Key	Points	Standard	Cluster	Explanation
13	MC	D	1	CCSS.MATH. CONTENT. 5.NF.B.7	Numbers and Operations Fractions	Three horses can eat one bale of hay. Since there are 6 bales, 18 horses can be fed.
14	MC	C	1	CCSS.MATH. CONTENT. 5.NBT.B.5	Numbers and Operations in Base Ten	Students can solve using models, place value strategies, or the traditional algorithm.
15	MC	C	1	CCSS.MATH. CONTENT. 5.G.A.1	Geometry	The first coordinate 6 is found on the x axis, the second coordinate 8 is found on the y axis. Students must recognize that in order to be a coordinate set, it must use parentheses.
16	MC	A	1	CCSS.MATH. CONTENT. 5.MD.C.4	Measurement and Data	To find the volume, count the width, length, and height. Multiply the three values.
17	MC	A	1	CCSS.MATH. CONTENT. 5.G.A.2	Geometry	Subtract the values of the Y coordinates. Chuck visited 12 cities and Danielle visited 4 cities. 12 - 4 = 8
18	MC	B	1	CCSS.MATH. CONTENT. 5.NF.B.5	Numbers and Operations Fractions	Students must recognize that $\frac{5}{5}$ is equal to 1 whole, so it would be the same as 713 x 1.
19	MC	C	1	CCSS.MATH. CONTENT. 5.NBT.B.6	Numbers and Operations in Base Ten	Students must use knowledge of place value to determine the equation most related to the scenario. Since there are 400 players being separated into 4 groups, they can relate the multiplication fact 4 x 100 = 400 to this scenario.
20	MC	B	1	CCSS.MATH. CONTENT. 5.MD.C.5.C	Measurement and Data	Multiply length x width x height to find the volume of each rectangular prism. Add them together.
21	MC	D	1	CCSS.MATH. CONTENT. 5.NBT.A.1	Numbers and Operations in Base Ten	The first two choices represent division, the second two represent multiplication. 50 x 100 = 5,000
22	MC	B	1	CCSS.MATH. CONTENT. 5.NBT.A.2	Numbers and Operations in Base Ten	Students must compare the value of the decimals first, then determine the value of the exponents. 4.86 is the larger of the two decimals. 10^{24} is the larger of the two exponents.
23	MC	A	1	CCSS.MATH. CONTENT. 5.NBT.B.7	Numbers and Operations in Base Ten	Students must add the times spent swimming laps. Common errors might be related to aligning the place values of decimals.

GO ON

Question	Type	Key	Points	Standard	Cluster	Explanation
24	MC	D	1	CCSS.MATH. CONTENT. 3.NBT.A.3	Numbers and Operations in Base Ten	Students must subtract their answer from the previous question to compare time spent swimming and running.
25	MC	B	1	CCSS.MATH. CONTENT. 5.NF.A.1	Numbers and Operations Fractions	Students must change denominators in order to add. The new denominator would become 24.
26	MC	C	1	CCSS.MATH. CONTENT. 5.NF.A.2	Numbers and Operations Fractions	Since $\frac{4}{5}$ is close to a whole, and $\frac{1}{4}$ is slightly larger than $\frac{1}{5}$, students can conclude that he used slightly more than 1 gallon of paint.
27	MC	D	1	CCSS.MATH. CONTENT. 5.NF.B.6	Numbers and Operations - Fractions	Students can multiply the numerators and denominators to find the answer. Common error might include adding the denominators or changing the denominators to be equal before multiplying the numerators.
28	MC	C	1	CCSS.MATH. CONTENT. 3.OA.A.2	Operations and Algebraic Thinking	Students must solve the expression in the parentheses first and then solve the expression from left to right.
29	MC	D	1	CCSS.MATH. CONTENT. 5.MD.C.3	Measurement and Data	This question will help determine whether students understand the meaning of volume and cubic units.
30	MC	A	1	CCSS.MATH. CONTENT. 5.MD.A.1	Measurement and Data	Multiply the number of gallons by 16 because there are 16 cups in a gallon.
Session 2						
31	MC	C	1	CCSS.MATH. CONTENT. 5.NF.B.6	Numbers and Operations Fractions	Multiply the number of blocks by the numerator. 8 x 3 is 24. Divide that by the denominator $\frac{24}{4}$ = 6.
32	MC	C	1	CCSS.MATH. CONTENT. 5.NBT.A.2	Numbers and Operations in Base Ten	There are two zeroes in 100, so the decimal will move to the right two times.
33	MC	A	1	CCSS.MATH. CONTENT. 5.NBT.A.4	Numbers and Operations in Base Ten	The numeral 8 is in the tenths place. There is a 7 in the hundredths place, indicating that the value of this number is closer to 9 tenths.
34	MC	A	1	CCSS.MATH. CONTENT. 5.NBT.B.5	Numbers and Operations in Base Ten	Students can solve using models, place value strategies, or the traditional algorithm.

Question	Type	Key	Points	Standard	Cluster	Explanation
35	MC	C	1	CCSS.MATH. CONTENT. 5.NBT.B.6	Numbers and Operations in Base Ten	Round the values in the scenario in order to determine which equation is closest.
36	MC	B	1	CCSS.MATH. CONTENT. 5.NBT.B.7	Numbers and Operations in Base Ten	Convert the numbers to standard form to compare.
37	MC	A	1	CCSS.MATH. CONTENT. 5.MD.A.1	Measurement and Data	There are 48 oz in three pounds. Divide that by 2 oz (needed by each student) to get 24.
38	MC	C	1	CCSS.MATH. CONTENT. 5.OA.A.2	Operations and Algebraic Thinking	Based on the phrasing, parentheses are implied because you have to find the sum of 18 and 3 before multiplying.
39	SR	Each will have 78 pieces of candy with 6 left over.	2	CCSS.MATH. CONTENT. 5.NBT.B.6	Numbers and Operations in Base Ten	Students can use a model or partial quotient to divide. They should show their work and determine the remainder.
40	SR	0.4; See Figure Below	2	CCSS.MATH. CONTENT. 5.NBT.B.7	Numbers and Operations in Base Ten	Students can divide by creating groups of two or by creating two groups. To earn both points, students must correctly draw model and identify the quotient.
41	ER	He is incorrect. He will need more than one quart of oil.	2	CCSS.MATH. CONTENT. 5.NF.A.2	Numbers and Operations Fractions	The first fraction, $\frac{3}{8}$, is close to $\frac{1}{2}$. The second fraction is more than $\frac{1}{2}$. Students can estimate and reason (without solving) that he will need more than one quart of oi.
42	SR	$\frac{1}{24}$; See Figure Below	2	CCSS.MATH. CONTENT. 5.NF.B.7	Numbers and Operations Fractions	Students must shade in the fraction $\frac{1}{4}$, then count the whole.
43	SR	See Figure Below	2	CCSS.MATH. CONTENT. 5.MD.B.2	Measurement and Data	To earn both points, students must correctly label the fractions on the number line. They also must plot the data points. Note that at least one of the denominators would need to be changed.
44	MC	A	1	CCSS.MATH. CONTENT. 5.NF.A.2	Numbers and Operations Fractions	Students need to subtract a fraction from a whole number. This will require them to 'break apart' the whole number to create an equivalent fraction.

GO ON

Question	Type	Key	Points	Standard	Cluster	Explanation
45	SR	See Figure Below	3	CCSS.MATH. CONTENT. 5.OA.B.3	Operations and Algebraic Thinking	To earn credit for three points students must correctly determine the coordinate points, plot them, and determine the rule.

Question	Figures

40

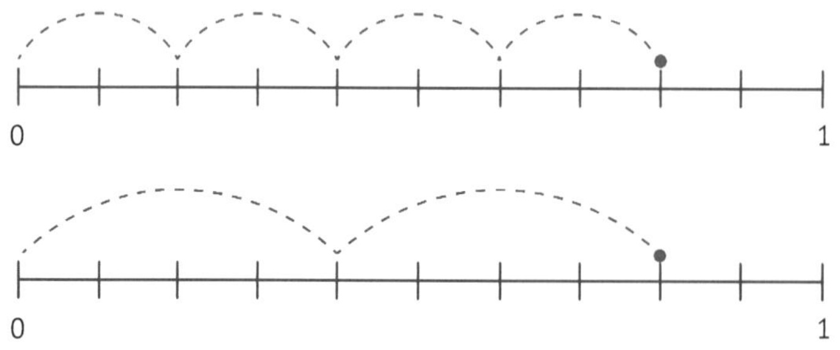

42

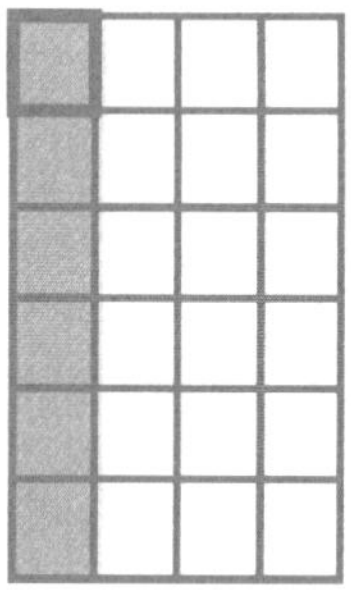

43

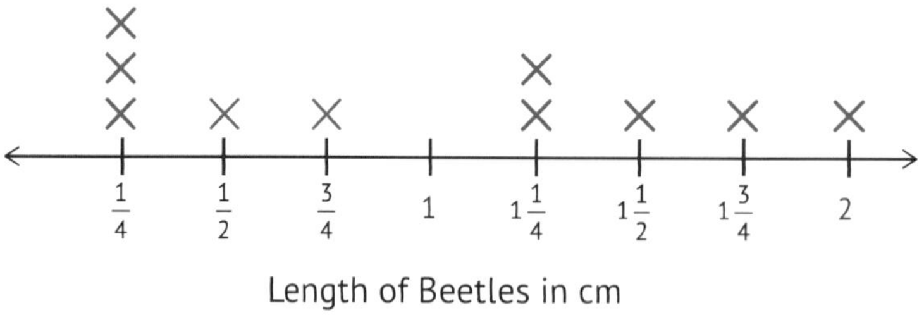

45

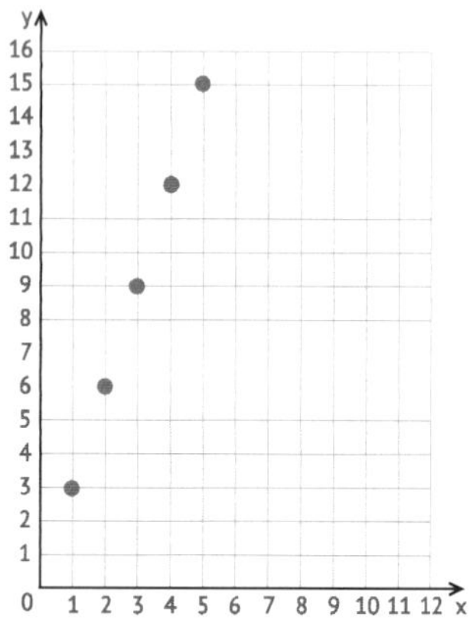

x	y	(x,y)
1	3	(1,3)
2	6	(2,6)
3	9	(3,9)
4	12	(4,12)
5	15	(5,15)
Rule = X × 3		

Question	Type	Key	Points	Standard	Cluster	Explanation
Session 1						
1	MC	D	1	CCSS.MATH. CONTENT. 5.NBT.B.6	Numbers and Operations in Base Ten	300 can be divided by 4 70 times with 20 leftover. Combine the 30 from the hundreds column with the 20 from the tens column to get 40. 40 divided by 4 is 10. 1 cannot be divided by 4 and so it is the remainder.
2	MC	C	1	CCSS.MATH. CONTENT. 5.NBT.B.5	Numbers and Operations in Base Ten	Students may use place value strategies or the traditional algorithm.
3	MC	D	1	CCSS.MATH. CONTENT. 5.NF.A.2	Numbers and Operations Fractions	Students can immediately reason that he ate more than $\frac{1}{2}$ pie because $\frac{5}{6}$ is close to a whole. If they solve the addition problem they would get the sum $\frac{26}{24}$ which is closest to 1 whole pie.
4	MC	D	1	CCSS.MATH. CONTENT. 5.G.B.3	Geometry	It is a quadrilateral because it has 4 sides. It is a parallelogram because it has one pair of parallel sides.
5	MC	D	1	CCSS.MATH. CONTENT. 5.NF.A.1	Numbers and Operations Fractions	When you change the denominator to 6 and add, you find the sum $\frac{7}{6}$. You need to change the improper fraction to a mixed number, $1\frac{1}{6}$, and then combine it with the whole numbers to get $8\frac{1}{6}$.
6	MC	B	1	CCSS.MATH. CONTENT. 5.NBT.B.7	Numbers and Operations in Base Ten	Students can use the visual model to sort the blocks into two different boxes. The flats represent whole dollars, the rods represent tenths, and the cubes represent hundredths.
7	MC	A	1	CCSS.MATH. CONTENT. 5.NBT.B.5	Numbers and Operations in Base Ten	Students may use place value strategies or the traditional algorithm.
8	MC	B	1	CCSS.MATH. CONTENT. 5.NF.B.4	Numbers and Operations Fractions	Students must multiply the numerators and denominators to find the area.

GO ON

Question	Type	Key	Points	Standard	Cluster	Explanation
9	MC	D	1	CCSS.MATH. CONTENT. 5.NF.B.5	Numbers and Operations in Base Ten	When students multiply the denominator they will divide each slice into 4 sections. They can then shade in 3 of those 4 pieces (ignoring the $\frac{1}{8}$th that wasn't shaded).
10	MC	C	1	CCSS.MATH. CONTENT. 5.NF.B.7	Numbers and Operations Fractions	Each box represents a tooth placed under the pillow. Students should draw two flats, two rods, and two units in each box. They should end up with 10 each.
11	MC	B	1	CCSS.MATH. CONTENT. 5.NBT.B.5	Numbers and Operations in Base Ten	Students must recognize that there is a zero in the product 10, and there are two additional zeroes in the product due to place value.
12	MC	B	1	CCSS.MATH. CONTENT. 5.NF.A.2	Numbers and Operations Fractions	Students can identify that $\frac{7}{8}$ is close to a whole, so they can estimate that the dog at 1 whole foot of the leash. That would leave $2\frac{1}{2}$ feet.
13	MC	C	1	CCSS.MATH. CONTENT. 5.NBT.B.6	Numbers and Operations in Base Ten	Students must recognize that the dividend 1,842 is closest to the product 1,800.
14	MC	A	1	CCSS.MATH. CONTENT. 5.NF.B.3	Numbers and Operations Fractions	Since there are not enough cakes for each family to get 1 whole, the cakes need to be split into 6ths. Each family would get $\frac{1}{6}$ from each of the cakes.
15	MC	D	1	CCSS.MATH. CONTENT. 5.NBT.A.1	Numbers and Operations in Base Ten	Since Ava sold less, the students can eliminate the first two answers. They must recognize that 5,000 is 100 times greater than 50, so 50 is $\frac{1}{100}$ the value of 5,000.
16	MC	C	1	CCSS.MATH. CONTENT. 5.G.B.4	Geometry	Because it needs to have two congruent sides and two congruent angles, all angles must be acute.
17	MC	C	1	CCSS.MATH. CONTENT. 5.MD.C.4	Measurement and Data	To find the volume, multiple length x height x width. 12 x 6 x 5 = 360 ft cubed

Question	Type	Key	Points	Standard	Cluster	Explanation
18	MC	C	1	CCSS.MATH. CONTENT. 5.MD.A.1	Measurement and Data	Either multiply 50 by 16 and determine the weight limit (800), then subtract. Divide the remainder by 16 to determine the number of lbs. OR Divide 896 by 16 to determine the number of lbs. Subtract 50 to determine how much the customer went over the limit.
19	MC	A	1	CCSS.MATH. CONTENT. 5.NBT.A.2	Numbers and Operations in Base Ten	The decimal in the answer,'moves' to the right two places. Students must recognize that the decimal shifts for every 0 in the factor. It shifts two places, so the missing factor must be 100.
20	MC	C	1	CCSS.MATH. CONTENT. 5.NBT.B.7	Numbers and Operations in Base Ten	Students can use a model or the traditional algorithm to solve.
21	MC	B	1	CCSS.MATH. CONTENT. 5.NF.B.6	Numbers and Operations Fractions	Students should shade the model to show $\frac{4}{5}$ in each one. They would add the numerator to find $\frac{24}{5}$. That can be simplified to $4\frac{4}{5}$.
22	MC	B	1	CCSS.MATH. CONTENT. 5.NF.B.5	Numbers and Operations Fractions	Students can reason that the correct answer is B by comparing the fractions to a whole number. $\frac{1}{5}$ is furthest from a whole number, $\frac{2}{3}$ is closer to a whole, $\frac{6}{6}$ is a whole number, and $\frac{3}{2}$ is greater than a whole.
23	MC	D	1	CCSS.MATH. CONTENT. 5.MD.C.5	Measurement and Data	Students must multiply the length, width, and height to find the volume of one box. They should multiply that area by 3 to represent the three boxes.
24	MC	C	1	CCSS.MATH. CONTENT. 5.NBT.A.4	Numbers and Operations in Base Ten	The 8 is in the hundredths place. The 9 in the thousandths place indicates that the weight is closer to 2.19 lbs than it is to 2.18 lbs.
25	MC	A	1	CCSS.MATH. CONTENT. 5.NF.B.7	Numbers and Operations Fractions	Students can use the model to cross of the amounts spent on concessions.

GO ON

Question	Type	Key	Points	Standard	Cluster	Explanation
26	MC	B	1	CCSS.MATH.CONTENT.5.MD.C.3	Measurement and Data	Count the cubic units to determine the width (3), length (6), and height (4). Multiply to find the volume.
27	MC	C	1	CCSS.MATH.CONTENT.5.MD.A.1	Measurement and Data	Multiply 3 times 16.
28	MC	C	1	CCSS.MATH.CONTENT.5.MD.B.2	Measurement and Data	This question is more about recognizing which information is important. To solve, it is not relevant that there are 3 runs that were 6 miles each. Instead, the students need to recognize that the longest run was 6 miles and the shortest run was $5\frac{1}{4}$ miles, then subtract.
29	MC	B	1	CCSS.MATH.CONTENT.5.NBT.A.3	Numbers and Operations in Base Ten	Since the word form ends in the thousandths place, the answer must have its last digit in the thousandths place.
30	MC	A	1	CCSS.MATH.CONTENT.5.MD.A.1	Measurement and Data	Multiply 128 by 2, then divide by 3.
Session 2						
31	MC	C	1	CCSS.MATH.CONTENT.5.NBT.B.6	Numbers and Operations in Base Ten	Students may use an area model or partial quotient strategy to solve.
32	MC	A	1	CCSS.MATH.CONTENT.5.NF.B.6	Numbers and Operations Fractions	48 times 4 would be 192. 192 divided by 3 is 64.
33	MC	Matt is correct.	2	CCSS.MATH.CONTENT.5.MD.C.3	Measurement and Data	In order to measure volume, the units of measurement need to be uniform. There also cannot be any overlap or gaps when physically using cubic units to measure volume
34	MC	A	1	CCSS.MATH.CONTENT.5.NBT.B.5	Numbers and Operations in Base Ten	Students should solve using the traditional algorithm
35	MC	D	1	CCSS.MATH.CONTENT.5.NBT.A.1	Numbers and Operations in Base Ten	The decimal would move to the right two places, adding two zeros to the figure. This increased value would make sense because they want to earn more money.

Question	Type	Key	Points	Standard	Cluster	Explanation
36	SR	He is not correct; $4\frac{3}{12}$ OR $4\frac{1}{4}$ inches	2	CCSS.MATH. CONTENT. 5.NF.A.1	Numbers and Operations Fractions	Students will need to solve by converting the denominator to 12 (and changing the numerators accordingly.)
37	MC	C	1	CCSS.MATH. CONTENT. 5.NF.B.7	Numbers and Operations Fractions	Students should draw five horizontal lines across to show the fraction being divided.
38	MC	A	1	CCSS.MATH. CONTENT. 5.MD.A.1	Measurement and Data	15 yards is 540 inches. 540 divided by 2 is 270.
39	SR	See figure below.	2	CCSS.MATH. CONTENT. 5.MD.B.2	Measurement and Data	Students need to number the line on the line plot before adding the data.
40	MC	B	2	CCSS.MATH. CONTENT. 5.OA.A.1	Operations and Algebraic Thinking	Students must solve the expressions in the parentheses first, then divide, then add.
41	MC	B	2	CCSS.MATH. CONTENT. 5.MD.A.1	Measurement and Data	To find the number of minutes multiply 2 times 60 = 120. To find the number of seconds multiply 120 times 60 = 720.
42	SR	She is incorrect. 8,925.	2	CCSS.MATH. CONTENT. 5.NBT.A.2	Numbers and Operations in Base Ten	Cassie is not moving the decimal when multiplying. She needs to move the decimal to the right twice because there are two zeros in 100.
43	SR	He is incorrect. 12,033	2	CCSS.MATH. CONTENT. 5.NBT.B.5	Numbers and Operations in Base Ten	Shawn forgot to indicate place value in the second partial product by adding a zero.
44	MC	D	1	CCSS.MATH. CONTENT. 5.NF.B.3	Numbers and Operations Fractions	There are 6 items to be divided among 7 people. Each roll would need to be divided into 7 pieces, then each person would take a piece from each of the 6 rolls.
45	SR	See figure below.	2	CCSS.MATH. CONTENT. 5.G.A.1	Geometry	The first number in the coordinate plane is a 5, which means you need to travel along the x axis to five. The second number is a 13, which means you need to travel along the y axis til you find the number 13. Trace the lines from both 5 and 13 to where they intersect. Draw a dot/point at that intersection. Label it (5, 13) - making sure to include parentheses.

GO ON ▶

<table>
<tr><th>Question</th><th>Figures</th></tr>
<tr><td>39</td><td>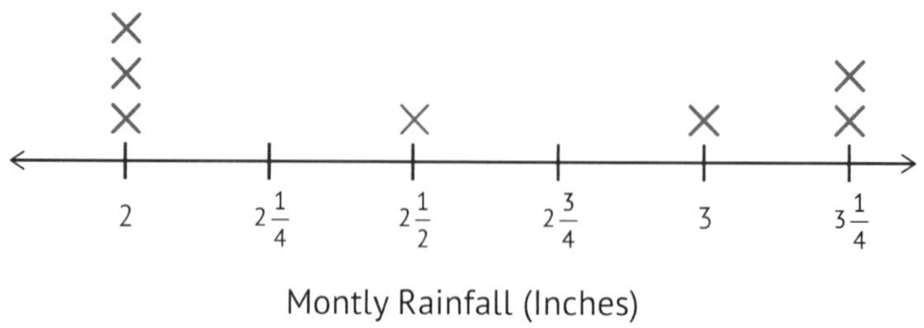</td></tr>
<tr><td>45</td><td>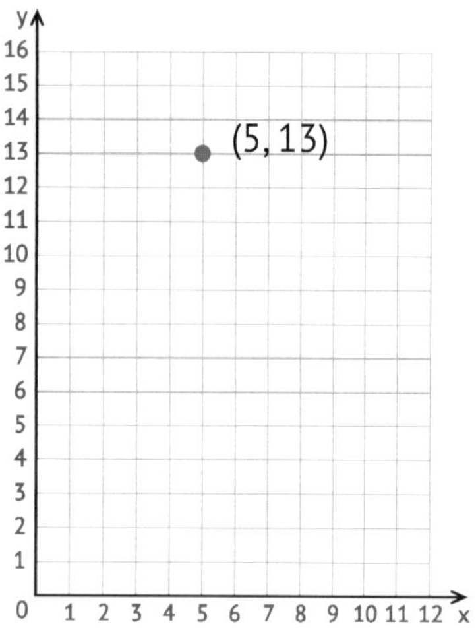</td></tr>
</table>

Question	Type	Key	Points	Standard	Cluster	Explanation
Session 1						
1	MC	B	1	CCSS.MATH.CONTENT.5.NBT.B.6	Numbers and Operations in Base Ten	Since the dividend is closest to 210, you can estimate that about 30 blankets would go in each box.
2	MC	B	1	CCSS.MATH.CONTENT.5.MD.A.1	Measurement and Data	Multiply 3 times 12 because there are 12 inches in one foot.
3	MC	D	1	CCSS.MATH.CONTENT.5.NF.A.2	Numbers and Operations Fractions	Both $\frac{3}{4}$ and $\frac{2}{3}$ are closer to 1 whole cup. If they are added to $1\frac{1}{2}$ cups, they yield $2\frac{11}{12}$, which is closest to 3 cups.
4	MC	A	1	CCSS.MATH.CONTENT.5.NBT.A.3	Numbers and Operations in Base Ten	When comparing decimals start with the greatest place value. If those digits match, continue looking to the right until you find a digit that matches. In this case, the hundredths place does not match. .01 is less than .04, so Matt won the race.
5	MC	C	1	CCSS.MATH.CONTENT.5.MD.B.2	Measurement and Data	Change the denominator of the mixed number in order to subtract. Subtract the fractions first. There is no other mixed number, so it is just a part of the difference.
6	MC	B	1	CCSS.MATH.CONTENT.5.MD.B.2	Measurement and Data	Add all of the data points from the line plot. Determine if there are any fractions that can add up to whole numbers, then add the fractions.
7	MC	C	1	CCSS.MATH.CONTENT.5.MD.B.2	Measurement and Data	$\frac{5}{6}$ hour is in the middle of the data set.
8	MC	C	1	CCSS.MATH.CONTENT.5.MD.A.1	Measurement and Data	Multiply 2.5 times 2,000 to get 5,000 lbs because 1 ton is equal to 2,000 lbs.
9	MC	A	1	CCSS.MATH.CONTENT.5.MD.C.4	Measurement and Data	Multiply the length, width, and height to get 6. Students must understand that it is measured in cubic feet rather than square feet because volume measures three dimensions.

GO ON

Question	Type	Key	Points	Standard	Cluster	Explanation
10	MC	A	1	CCSS.MATH. CONTENT. 5.NBT.B.7	Numbers and Operations in Base Ten	The rods and cubes represent the change. Rods are dimes and cubes are pennies. (8 + 1 = 9 ; 4 + 5 = 9). The flats represent whole dollars and can be combined to make 5.
11	MC	B	1	CCSS.MATH. CONTENT. 5.NF.A.1	Numbers and Operations Fractions	In order to solve, change the denominator to 12 add fractions first. When you get the sum of $\frac{17}{12}$. Regroup to make a mixed number, then add to the whole numbers.
12	MC	A	1	CCSS.MATH. CONTENT. 5.NF.B.6	Numbers and Operations Fractions	Change the mixed number to an improper fraction, then multiply to get $\frac{20}{2}$. Simplify to get 10.
13	MC	C	1	CCSS.MATH. CONTENT. 5.NBT.B.5	Numbers and Operations in Base Ten	Students should solve using the traditional algorithm.
14	MC	A	1	CCSS.MATH. CONTENT. 5.MD.C.5	Measurement and Data	Students can solve by multiplying all of the options, or they could reason that 10 times 10 is 100, which would be more than 1,000 when multiplied with a two-digit number.
15	MC	B	1	CCSS.MATH. CONTENT. 5.NBT.A.1	Numbers and Operations in Base Ten	When comparing with a fractional amount you divide by the denominator. To divide by 100, move the decimal to the left two place value.
16	MC	B	1	CCSS.MATH. CONTENT. 5.NBT.B.6	Numbers and Operations in Base Ten	Seven times 9 is 63, which can be taken from 70.
17	MC	B	1	CCSS.MATH. CONTENT. 5.NF.B.7	Numbers and Operations Fractions	$\frac{1}{4}$ split into 4 pieces would be $\frac{1}{16}$ of the original cake.
18	MC	C	1	CCSS.MATH. CONTENT. 5.G.B.4	Geometry	Four and twenty should be in parenthesis because the sum needs to be determined before it is divided by six.
19	MC	C	1	CCSS.MATH. CONTENT. 5.NF.B.7	Numbers and Operations Fractions	$\frac{1}{2}$ split into 5 pieces would be $\frac{1}{10}$ of the original gallon.

Question	Type	Key	Points	Standard	Cluster	Explanation
20	MC	A	1	CCSS.MATH. CONTENT. 5.OA.B.3	Operations and Algebraic Thinking	Add five to the X coordinate.
21	MC	A	1	CCSS.MATH. CONTENT. 5.MD.C.3	Measurement and Data	Students must reason that a container used to store cereal would be smaller than a meter.
22	MC	B	1	CCSS.MATH. CONTENT. 5.NBT.B.6	Numbers and Operations Base Ten	When dividing 227 by three, you can make at least 70 groups because 7 times 3 is 21.
23	MC	C	1	CCSS.MATH. CONTENT. 5.NBT.A.2	Numbers and Operations in Base Ten	When multiplying a decimal, you move the decimal to the right based on the exponent. This equation would require two shifts to the right, which would be equal to multiplying it by 100.
24	MC	C	1	CCSS.MATH. CONTENT. 5.NBT.A.4	Numbers and Operations in Base Ten	First, identify the digit in the tenths place (7). Look at the digit to the right. Since it is a 1 (equal to 1 hundredth) it indicates that the number is closer to 5.7 than to 5.8.
25	MC	A	1	CCSS.MATH. CONTENT. 5.NBT.B.5	Numbers and Operations in Base Ten	Students should solve using the traditional algorithm.
26	MC	D	1	CCSS.MATH. CONTENT. 5.NBT.B.7	Numbers and Operations in Base Ten	Students should draw base ten blocks representing $4.25 in each of the squares, then add them.
27	MC	D	1	CCSS.MATH. CONTENT. 5.NF.A.2	Numbers and Operations Fractions	Change the denominator of the $\frac{1}{3}$ fraction to match the denominator of 9. Add and get $\frac{10}{9}$ gallons. Change to a mixed number to get $1\frac{1}{9}$ gallons.
28	MC	B	1	CCSS.MATH. CONTENT. 5.NF.B.3	Numbers and Operations Fractions	Students need to recognize that since there are fewer pieces than cats, each cat would not get a whole piece of cheese. The number of pieces is the numerator, the number of cats is the denominator.

GO ON

Question	Type	Key	Points	Standard	Cluster	Explanation
29	MC	C	1	CCSS.MATH. CONTENT. 5.NF.B.4	Numbers and Operations Fractions	Multiply the numerator times the whole number to get $\frac{9}{4}$. Simplify to get $2\frac{1}{4}$ candy bars.
30	MC	A	1	CCSS.MATH. CONTENT. 5.NF.B.5	Numbers and Operations Fractions	Students can reason that the product is greater than 42 because the fraction is greater than one whole.
Session 2						
31	MC	D	1	CCSS.MATH. CONTENT. 5.NF.A.2	Numbers and Operations Fractions	Change the denominator to 6 in order to add. Add the fractions first to get $\frac{9}{6}$. Change the improper fraction to a mixed number $1\frac{3}{6}$. Add to the whole numbers to get $3\frac{3}{6}$. Simplify to get $3\frac{1}{2}$.
32	MC	C	1	CCSS.MATH. CONTENT. 5.MD.A.1	Measurement and Data	Multiply 12 times 16 because there are 16 ounces in one pound.
33	SR	The missing data points are 16 and 6. The rule is $\frac{x}{2}$.	2	CCSS.MATH. CONTENT. 5.OA.A.1	Operations and Algebraic Thinking	To determine the rule, examine the relationship between x and y data points. It's clear that they are decreasing in value, so we can determine that the rule will be to either subtract or divide. The first data set decreases by 12, then the next data set decreases by 10, so there is not a consistent subtraction rule. Instead, consider division. 24 divided by something is 12. This missing divisor would be 2. If you check with another data set the rule is consistent 20 divided by 2 is 10.
34	MC	C	1	CCSS.MATH. CONTENT. 5.NF.B.4	Numbers and Operations Fractions	Multiply the numerator times the whole number. Get an improper fraction $\frac{14}{5}$. Changed the improper fraction to a mixed number $2\frac{4}{5}$.

Question	Type	Key	Points	Standard	Cluster	Explanation
35	SR	<	2	CCSS.MATH. CONTENT. 5.NBT.A.3	Numbers and Operations in Base Ten	When comparing numbers, start comparing at the greatest place value. If the values match, move to a lesser place value. In this case, the digits match until the thousandths place. One number doesn't have a digit in the thousandths place, so it can be thought of as zero. 2 > 0
36	ER	length, height, width	2	CCSS.MATH. CONTENT. 5.MD.C.3	Measurement and Data	Laura needs to determine the volume of each box. In order to do so she needs to measure the length, width, and height. She should multiply those dimensions to determine how much space is inside the box. This will help her determine how much can be used to fill the box.
37	SR	9	2	CCSS.MATH. CONTENT. 5.NF.B.5	Numbers and Operations Fractions	Separate the shapes into four groups. Shade three of the four groups. See figure.
38	SR	multiplication	2	CCSS.MATH. CONTENT. 5.MD.A.1	Measurement and Data	The teacher needs to multiply. There are 16 cups in a gallon, and there are three gallons. That means there are three groups of 16, which would require multiplication.
39	SR	right triangle	2	CCSS.MATH. CONTENT. 5.G.A.1	Geometry	This is a right triangle because one of the angles is 90 degrees. It is a scalene triangle because none of the sides are even.
40	MC	C	1	CCSS.MATH. CONTENT. 5.NF.A.1	Numbers and Operations Fractions	Students can use the traditional algorithm or the visual model to help illustrate 8.65 multiplied by 12.
41	MC	B	1	CCSS.MATH. CONTENT. 5.NBT.B.6	Numbers and Operations in Base Ten	The missing partial quotient is 50 because 4 times 50 is 200.
42	MC	B	1	CCSS.MATH. CONTENT. 5.NBT.B.5	Numbers and Operations in Base Ten	Students can either test each equation or estimate to see which is closest to the product.
43	MC	C	1	CCSS.MATH. CONTENT. 5.NBT.A.4	Numbers and Operations in Base Ten	Students must recognize that the digit 1 is in the hundredths place. Since there is a 9 in the thousandths place, they can reason that it is closer to 2 hundredths than 1 hundredth.

GO ON

Question	Type	Key	Points	Standard	Cluster	Explanation
44	SR	1,176	2	CCSS.MATH. CONTENT. 5.NBT.B.6	Numbers and Operations in Base Ten	She forgot to regroup when finding the first partial product. She created three new tens when she multiplied 9 times 4.
45	SR	10	2	CCSS.MATH. CONTENT. 5.OA.A.1	Operations and Algebraic Thinking	He did not solve it correctly. He performed the operations in order from left to right. He should have solved the portion in the parentheses first. After solving the portion in the parentheses he should have divided. Finally, he should have added.

Question	Figures
37	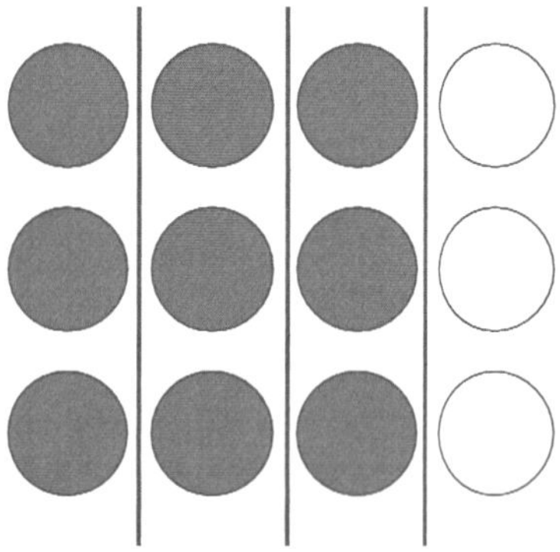
45	4 + (3 x 12) ÷ 6 4 + 36 ÷ 6 4 + 6 10

Question	Type	Key	Points	Standard	Cluster	Explanation
Session 1						
1	MC	A	1	CCSS.MATH. CONTENT. 5.NBT.B.5	Numbers and Operations in Base Ten	Students should solve using the traditional algorithm.
2	MC	C	1	CCSS.MATH. CONTENT. 5.NBT.A.4	Numbers and Operations in Base Ten	Each tally between the numbers represents one hundredth, so the point represents 7.37. It is closer to 7.4, so it rounds tp 7.4.
3	MC	D	1	CCSS.MATH. CONTENT. 5.NF.B.7	Numbers and Operations Fractions	Draw three whole shapes to represent the three gallons, divide each into two parts. Count the number of parts.
4	MC	B	1	CCSS.MATH. CONTENT. 5.NF.B.4	Numbers and Operations Fractions	Two out of three rows are shaded. Five out of six columns are shaded. The matching equation is $\frac{2}{3} \times \frac{5}{6}$.
5	MC	A	1	CCSS.MATH. CONTENT. 5.MD.C.4	Measurement and Data	Count the cubic units to determine the length (3), width (2), and height (3). Multiply all three to find the volume.
6	MC	B	1	CCSS.MATH. CONTENT. 5.NF.B.7	Numbers and Operations Fractions	Draw five whole shapes to represent the flower beds. Divide each into three parts. Count the number of parts.
7	MC	C	1	CCSS.MATH. CONTENT. 5.NF.A.2	Numbers and Operations Fractions	Change the denominator to 12. When you add the fractions, they yield $\frac{27}{12}$. That can be turned in $2\frac{3}{12}$. If you add that to the 2 whole numbers, that is $4\frac{3}{12}$. Since $\frac{3}{12}$ is closer to 0 than it is to a whole, round down. It is closer to 4 hours than to 5.
8	MC	B	1	CCSS.MATH. CONTENT. 5.MD.B.2	Measurement and Data	Count the number of Xs that sits at 18.75 on the number line.
9	MC	C	1	CCSS.MATH. CONTENT. 5.NBT.B.5	Numbers and Operations in Base Ten	Students should solve using the traditional algorithm.
10	MC	C	1	CCSS.MATH. CONTENT. 5.MD.A.1	Measurement and Data	Multiply 20 times 3 because there are three feet in one yard.

GO ON

Question	Type	Key	Points	Standard	Cluster	Explanation
11	MC	D	1	CCSS.MATH. CONTENT. 5.NBT.B.5	Numbers and Operations in Base Ten	Students should solve using the traditional algorithm.
12	MC	B	1	CCSS.MATH. CONTENT. 5.NF.B.5	Numbers and Operations Fractions	Students can reason that B is greater than 724 because the fraction represents more than 1 whole.
13	MC	C	1	CCSS.MATH. CONTENT. 5.NF.A.2	Numbers and Operations Fractions	Students can create a visual model to show the yard with a denominator of 12. They can also change the denominator to 12, then add. They get the sum $\frac{7}{12}$, which means more than $\frac{4}{12}$ $\left(\text{or } \frac{1}{4}\right)$ remains.
14	MC	C	1	CCSS.MATH. CONTENT. 5.NF.B.7	Numbers and Operations Fractions	$\frac{1}{2}$ divided into 6 parts would be $\frac{1}{12}$ of the original whole cup.
15	MC	D	1	CCSS.MATH. CONTENT. 5.NF.B.3	Numbers and Operations Fractions	Students should reason that since the monkeys outnumber bananas, none of them can have a whole banana. The zookeeper would have to cut each banana into 12 pieces, then give $\frac{1}{12}$ to each monkey. The monkeys would get $\frac{11}{12}$ because there are 11 bananas.
16	MC	C	1	CCSS.MATH. CONTENT. 5.NBT.B.7	Numbers and Operations in Base Ten	Students will cross off base ten blocks to show money 'spent.' They will have to regroup and break apart blocks in order to subtract correctly.
17	MC	B	1	CCSS.MATH. CONTENT. 5.NF.B.6	Numbers and Operations Fractions	Change the mixed number to an improper fraction. Multiply by whole number to get $\frac{110}{3}$. Simplify to get $36\frac{2}{3}$.
18	MC	B	1	CCSS.MATH. CONTENT. 5.NF.A.2	Numbers and Operations Fractions	Students can use the visual model to subtract. They will subtract a total of $\frac{5}{10}$ from the $\frac{8}{10}$ to get $\frac{3}{10}$.

Question	Type	Key	Points	Standard	Cluster	Explanation
19	MC	C	1	CCSS.MATH.CONTENT.5.NF.B.6	Numbers and Operations Fractions	Change the mixed number to an improper fraction. Multiply to get the product $\frac{60}{2}$. Simplify/divide to get 30.
20	MC	A	1	CCSS.MATH.CONTENT.5.MD.C.3	Measurement and Data	Students must reason that a box used to store Legos would be smaller than yards, meters, or kilometers.
21	MC	B	1	CCSS.MATH.CONTENT.5.MD.A.1	Measurement and Data	Multiply 5 times 16 because there are 16 ounces in one pound.
22	MC	B	1	CCSS.MATH.CONTENT.5.NF.B.7	Numbers and Operations Fractions	Students must know that cubic units are 'placed' inside a shape to determine how much would fill it.
23	MC	C	1	CCSS.MATH.CONTENT.5.NBT.B.6	Numbers and Operations in Base Ten	8 goes into 390 four times. Then subtract 320 to get 71. 71 divided by 8 is 8 with a remainder of 7.
24	MC	C	1	CCSS.MATH.CONTENT.5.NBT.A.1	Numbers and Operations in Base Ten	650,000 is ten times greater than 65,000.
25	MC	A	1	CCSS.MATH.CONTENT.5.G.A.1	Geometry	Coordinate pairs must be in parentheses. The first one listed must be found along the x axis. The second one must be along the y axis.
26	MC	B	1	CCSS.MATH.CONTENT.5.NBT.B.6	Numbers and Operations in Base Ten	The dividend is 200 than it is to 250, so when estimating she wants her product to be closer to 200 than to 250.
27	MC	A	1	CCSS.MATH.CONTENT.5.NF.A.1	Numbers and Operations Fractions	Change the denominator to be 8, then subtract fractions first.
28	MC	D	1	CCSS.MATH.CONTENT.5.MD.C.4	Measurement and Data	Multiply the length, width, and height. Students must recognize that it is measured in cubic units because it is measuring three dimensions.
29	MC	B	1	CCSS.MATH.CONTENT.5.MD.C.5	Measurement and Data	Multiply the dimensions of the two boxes, then add.
30	MC	A	1	CCSS.MATH.CONTENT.5.MD.A.1	Measurement and Data	Divide 4,521 by 100 and move the decimal two places to the left.

GO ON ▶

Question	Type	Key	Points	Standard	Cluster	Explanation
Session 2						
31	SR	See figure.	1	CCSS.MATH. CONTENT. 5.G.A.1	Geometry	See figure.
32	MC	C	1	CCSS.MATH. CONTENT. 5.NBT.A.1	Numbers and Operations in Base Ten	When dividing by one hundred, move the decimal two places to the left.
33	SR	See figure.	2	CCSS.MATH. CONTENT. 5.MD.B.2	Measurement and Data	Find the greatest and least values and create a number line which increases by $\frac{1}{4}$ each time.
34	SR	9,472	2	CCSS.MATH. CONTENT. 5.NBT.B.5	Numbers and Operations in Base Ten	He forgot to regroup when adding the partial products. He made a new group of one thousand when adding the hundreds digits and forgot to include it when adding the thousands digits.
35	MC	B	1	CCSS.MATH. CONTENT. 5.MD.C.4	Measurement and Data	Solve either by testing out each measurement, or multiply 6 times 7 to get 42. Divide 168 by 42 to get 4.
36	MC	D	1	CCSS.MATH. CONTENT. 5.NBT.A.3	Numbers and Operations in Base Ten	There are no numbers representing the thousands or tenths place, so these are represented with zeroes.
37	SR	$2\frac{2}{6}$ OR $2\frac{1}{3}$	2	CCSS.MATH. CONTENT. 5.NF.A.1	Numbers and Operations Fractions	Change the denominator of the fraction with the denominator of 3 by dividing each piece into two. See figure.
38	MC	B	1	CCSS.MATH. CONTENT. 5.NF.B.6	Numbers and Operations Fractions	Change the mixed number to an improper fraction $\left(\frac{11}{6}\right)$. Multiply the improper fraction by $\frac{2}{3}$ to get the product $\frac{22}{18}$. Change to a mixed number $\left(1\frac{4}{18}\right)$. Simplify to get $1\frac{2}{9}$.
39	ER	$(6 \times 8) - (15 \div 5)$; 45	3	CCSS.MATH. CONTENT. 5.OA.A.2	Operations and Algebraic Thinking	Willow did not group the numbers. She multiplied 6 times 8 and got 48. She saw the word minus and subtracted 15, then divided by 6.

Question	Type	Key	Points	Standard	Cluster	Explanation
40	SR	It is a parallelogram. It is a quadrilateral. It is a rectangle.	2	CCSS.MATH. CONTENT. 5.G.B.3	Geometry	All rectangles are both parallelograms and quadrilaterals.
41	MC	B	1	CCSS.MATH. CONTENT. 5.NF.A.2	Numbers and Operations Fractions	Change the denominator to four in order to add. Add the fractions and get $\frac{5}{4}$. Change to a mixed number $\left(1\frac{1}{4}\right)$. Add the whole numbers to get $10\frac{1}{4}$.
42	SR	Incorrect	1	CCSS.MATH. CONTENT. 5.MD.A.1	Measurement and Data	He divided by 1,000 when he should have multiplied by 1,000. It does not make sense to determine there is only about half a meter, since there are 1,000 meters in a kilometer.
43	SR	$2\frac{2}{12}$ gallons or $2\frac{1}{6}$ gallons	2	CCSS.MATH. CONTENT. 5.NF.A.2	Numbers and Operations Fractions	See figure.
44	MC	A	2	CCSS.MATH. CONTENT. 5.NBT.B.6	Numbers and Operations in Base Ten	Students should find partial quotients using the area model.
45	MC	A	2	CCSS.MATH. CONTENT. 5.NBT.B.5	Numbers and Operations in Base Ten	Students should solve using the traditional algorithm.

Question	Figures
31	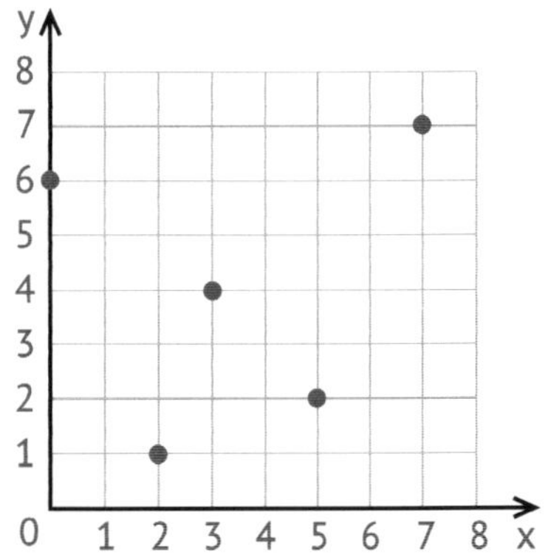

GO ON

Question	Figures
33	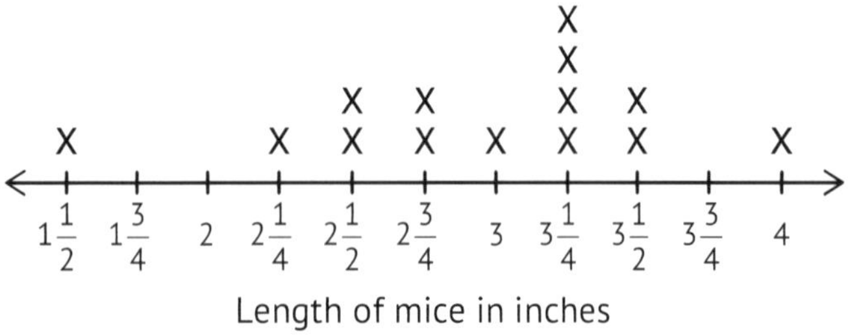
37	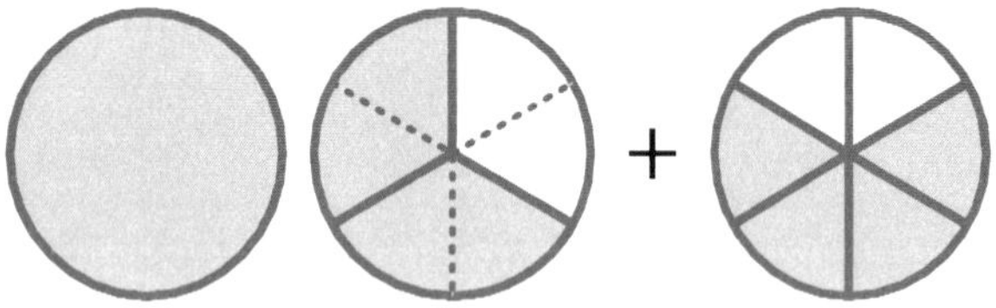 $1\frac{2}{3} + \frac{4}{6}$ $1\frac{4}{6} + \frac{4}{6}$ $1\frac{8}{6} \rightarrow 2\frac{2}{6}$
43	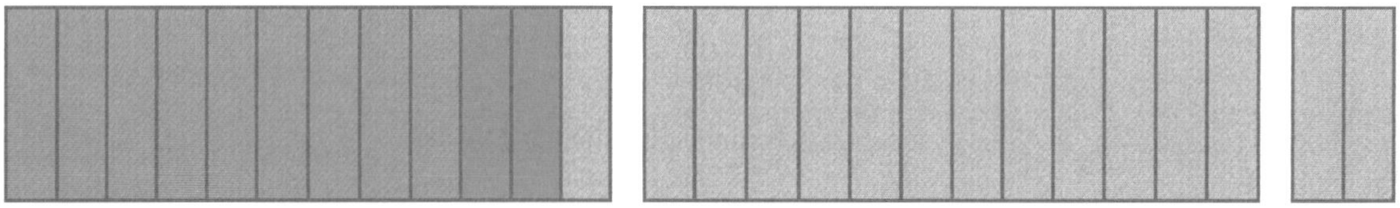  $\frac{3}{4} + 1\frac{1}{4} + \frac{1}{6}$ $\frac{9}{12} + 1\frac{3}{12} + \frac{2}{12} = 1\frac{14}{12} \rightarrow 2\frac{2}{12}$, or $2\frac{1}{6}$

Question	Type	Key	Points	Standard	Cluster	Explanation
Session 1						
1	MC	C	1	CCSS.MATH. CONTENT. 5.NBT.B.6	Numbers and Operations in Base Ten	The product of the multiplication equation needs to be less than the divisor. 720 is as close as these multiples of 12 will get without going over.
2	MC	B	1	CCSS.MATH. CONTENT. 5.NBT.B.6	Numbers and Operations in Base Ten	Using the answer from the previous question, students know that the quotient would be close to 60. They can subtract 720 from the number of campers and get 24. That would add 2 more to the quotient, so the quotient would be 62.
3	MC	D	1	CCSS.MATH. CONTENT. 5.NF.B.6	Numbers and Operations Fractions	Multiply the whole number by the numerator to get 32. 32 divided by 5 is 6 remainder 2, so the leash is between 6 and 7 meters long.
4	MC	D	1	CCSS.MATH. CONTENT. 5.MD.A.1	Measurement and Data	Multiply the number of tablespoons per cup (16) by the number of cups (4).
5	MC	A	1	CCSS.MATH. CONTENT. 5.NBT.A.3	Numbers and Operations in Base Ten	The standard form of the expanded number is 547.52. The only option that is greater is 547.53, which is .01 greater.
6	MC	B	1	CCSS.MATH. CONTENT. 5.NF.B.4	Numbers and Operations Fractions	The blue rows represent $\frac{2}{4}$. The yellow column represents $\frac{1}{5}$. The green shaded portion is the numerator of the product. Count all of the pieces as the denominator for the product.
7	MC	B	1	CCSS.MATH. CONTENT. 5.NBT.B.5	Numbers and Operations in Base Ten	Students should solve using the traditional algorithm.
8	MC	C	1	CCSS.MATH. CONTENT. 5.NF.A.2	Numbers and Operations Fractions	Change the denominator of the fractions to 10. Add $\frac{5}{10} + \frac{2}{10} + \frac{3}{10} + \frac{1}{10}$ to get $\frac{11}{10}$. Change to a mixed number $1\frac{1}{10}$ and combine with the whole number 1 to get $2\frac{1}{10}$.

GO ON ▶

Question	Type	Key	Points	Standard	Cluster	Explanation
9	MC	C	1	CCSS.MATH. CONTENT. 5.MD.C.3	Measurement and Data	The formula for volume is length x width x height. Weight is not a factor.
10	MC	A	1	CCSS.MATH. CONTENT. 5.OA.B.3	Operations and Algebraic Thinking	This is a list of multiples of 9. Each time a new group of 9 is taken away, so the rule is to subtract 9 each time.
11	MC	C	1	CCSS.MATH. CONTENT. 5.NBT.B.5	Numbers and Operations in Base Ten	Students should solve using the traditional algorithm.
12	MC	B	1	CCSS.MATH. CONTENT. 5.NBT.A.2	Numbers and Operations in Base Ten	The expression with the greatest exponent will yield the greatest product.
13	MC	A	1	CCSS.MATH. CONTENT. 5.NF.B.3	Numbers and Operations Fractions	The number of cups would be the numerator because that is what is being divided. The cups would be split into 6 pieces, so that would be the denominator.
14	MC	C	1	CCSS.MATH. CONTENT. 5.NBT.A.1	Numbers and Operations in Base Ten	When multiplying by 100 you move the decimal two times to the right.
15	MC	A	1	CCSS.MATH. CONTENT. 5.MD.C.4	Measurement and Data	Count the length (3), width (3), and height (3). Multiply the three dimensions to determine the volume.
16	MC	B	1	CCSS.MATH. CONTENT. 5.MD.B.2	Measurement and Data	He measured the wingspans of 10 birds because there are 10 Xs on the line plot.
17	MC	C	1	CCSS.MATH. CONTENT. 5.MD.B.2	Measurement and Data	The largest wingspan is $17\frac{1}{2}$ inches. The smallest wingspan is $16\frac{1}{2}$. The difference between them is 1 inch.
18	MC	C	1	CCSS.MATH. CONTENT. 5.NBT.B.6	Numbers and Operations in Base Ten	When you subtract and get 14, move it to the next column. Eight goes into 14 two times, so the next partial quotient is 1.
19	MC	A	1	CCSS.MATH. CONTENT. 5.NF.B.5	Numbers and Operations Fractions	The fraction $\frac{2}{3}$ is less than 1 whole. All of the other fractions are greater than or equal to one whole, so they would generate products that are greater than the factor 156.
20	MC	C	1	CCSS.MATH. CONTENT. 5.MD.C.5	Measurement and Data	Multiply the length, width and height to get 120. Since each dice is one cubic centimeter, he can fit 120 dice in the box.

Question	Type	Key	Points	Standard	Cluster	Explanation
21	MC	B	1	CCSS.MATH. CONTENT. 5.OA.A.1	Operations and Algebraic Thinking	Solve the expression in the parentheses first to get 17. Multiply 17 times 8 to get 136. Add 5 to get 141.
22	MC	C	1	CCSS.MATH. CONTENT. 5.NBT.A.4	Numbers and Operations in Base Ten	Each of the tally marks on the number line represents values in the hundredths place. The point is on 48.58 and is closest to 48.6
23	MC	B	1	CCSS.MATH. CONTENT. 5.NF.A.1	Numbers and Operations Fractions	Change the denominator to 10 by multiplying both the numerator and denominator by 2. Add $\frac{6}{10}$ plus $\frac{3}{10}$ to get $\frac{9}{10}$.
24	MC	A	1	CCSS.MATH. CONTENT. 5.NBT.B.7	Numbers and Operations in Base Ten	Subtract 72.45-45.92 to determine the amount spent at the store.
25	MC	A	1	CCSS.MATH. CONTENT. 5.NF.B.6	Numbers and Operations Fractions	Multiply the numerators to get 6. Multiply the denominators to get 24. $\frac{6}{24}$ can be simplified by dividing by 6 to get $\frac{1}{4}$.
26	MC	D	1	CCSS.MATH. CONTENT. 5.NF.B.7	Numbers and Operations Fractions	Each pie would have 8 pieces, and there are 4 pies. There would be 32 pieces total.
27	MC	A	1	CCSS.MATH. CONTENT. 5.MD.C.5	Measurement and Data	Find the volume of one box (4), then multiply by the total number of boxes (16).
28	MC	B	1	CCSS.MATH. CONTENT. 5.G.A.2	Geometry	The x axis represents the number of minutes played. The y axis represents the number of tickets earned. The point on the coordinate plane representing 8 minutes played also represents 7 tickets earned.
29	MC	C	1	CCSS.MATH. CONTENT. 5.NF.B.6	Numbers and Operations Fractions	Multiply the whole number by the numerator to get 18. 18 divided by 4 is 4 remainder 2, so the friends can buy 4 pizzas.
30	MC	B	1	CCSS.MATH. CONTENT. 5.NF.B.7	Numbers and Operations Fractions	The blue shaded region represents $\frac{1}{10}$. Each tenth is divided into 3 sections. The boxed portion of the shaded region represents the quotient. It is 1 out of 30 pieces or $\frac{1}{30}$th.

GO ON

Question	Type	Key	Points	Standard	Cluster	Explanation
Session 2						
31	MC	B	1	CCSS.MATH. CONTENT. 5.NF.B.6	Numbers and Operations Fractions	Change the mixed number to an improper fraction and multiply by the whole number. Divide by the denominator to create another mixed number. The remainder is the numerator.
32	MC	B	1	CCSS.MATH. CONTENT. 5.NBT.B.5	Number and Operations in Base Ten	Students should solve using the traditional algorithm.
33	ER	$\frac{\frac{1}{4}}{3}$	2	CCSS.MATH. CONTENT. 5.NF.B.7	Numbers and Operations Fractions	Answers may vary. Sample response: There is only $\frac{1}{4}$ of a pizza left. 3 friends decide to split it evenly. How much pizza would each friend get?
34	SR	$\frac{23}{24}$	2	CCSS.MATH. CONTENT. 5.NF.A.1	Numbers and Operations Fractions	Shade one circle to show $\frac{14}{24}$. Shade the other circle to show $\frac{3}{8}$. Change the denominator of the circle with 8ths by dividing each piece into three pieces, creating a denominator of 24. The shaded region now shows $\frac{9}{24}$. Add $\frac{9}{24}$ to $\frac{14}{24}$ and get $\frac{23}{24}$.
35	MC	D	1	CCSS.MATH. CONTENT. 5.MD.A.1	Measurement and Data	Multiply 12 by 12 to show the number of fluid oz available. Divide that by 8 to represent the capacity of each cup.
36	SR	129; See figure below.	2	CCSS.MATH. CONTENT. 5.NBT.B.6	Number and Operations in Base Ten	129; See figure below.
37	SR	See figure below.	2	CCSS.MATH. CONTENT. 5.MD.B.2	Measurement and Data	Create a number line that inclues 16 through $17\frac{1}{2}$. Draw tally marks to divide each whole number into fourths and label. Then divide each fourth into two sections to create eighths, then label. Add and X above a number for each time it occurs in the data set.

Question	Type	Key	Points	Standard	Cluster	Explanation
38	MC	C	1	CCSS.MATH. CONTENT. 5.NBT.B.6	Number and Operations in Base Ten	The dividend is closest to 800, and the divisor is 40, so she can estimate that she would have 200 ounces per flower pot.
39	SR	$4\frac{7}{12}$; See figure below.	2	CCSS.MATH. CONTENT. 5.NF.A.2	Numbers and Operations Fractions	See figure. Note: Students may also choose to represent fractions with circles or a number line.
40	MC	C	1	CCSS.MATH. CONTENT. 5.MD.C.3	Measurement and Data	Volume means to determine the capacity of cubic units in a space. Determining how much space is in a box would require measurements of length, width, and height- the dimensions of volume.
41	SR	$\frac{1}{8}$; See figure below.	2	CCSS.MATH. CONTENT. 5.NF.B.6	Numbers and Operations Fractions	See figure. Divide the wall into two pieces and shade one side to represent Thomas' portion. Divide that horizontally into 4 sections and shade one to represent the amount he has completed so far. The overlapping portion is the numerator and the entire figure represents the denominator.
42	MC	A	1	CCSS.MATH. CONTENT. 5.NF.B.5	Numbers and Operations Fractions	$\frac{4}{3}$ and $\frac{3}{2}$ are greater than 1 whole. $\frac{7}{7}$ is equal to 1 whole. $\frac{1}{2}$ is less than 1 whole, so it would yield the smallest product.
43	MC	C	2	CCSS.MATH. CONTENT. 5.NF.B.4	Numbers and Operations Fractions	Change the mixed number to an improper fraction $\left(\frac{29}{4}\right)$. Multiply by the whole number (18) to get $\frac{522}{4}$. 522 divided by 4 is 130 R 2, or $130\frac{2}{4}$, or $130\frac{1}{2}$.
44	SR	32,000	2	CCSS.MATH. CONTENT. 5.NBT.A.2	Number and Operations in Base Ten	When multiplying by multiples of ten the number of zeros in a factor will be equal to the number of zeros in a product.
45	MC	D	2	CCSS.MATH. CONTENT. 5.OA.A.2	Operations and Algebraic Thinking	Since the phrase 'the quotient' is used, it indicates that division should happen first, and should therefore be put in parentheses.

GO ON

Question	Figures
34	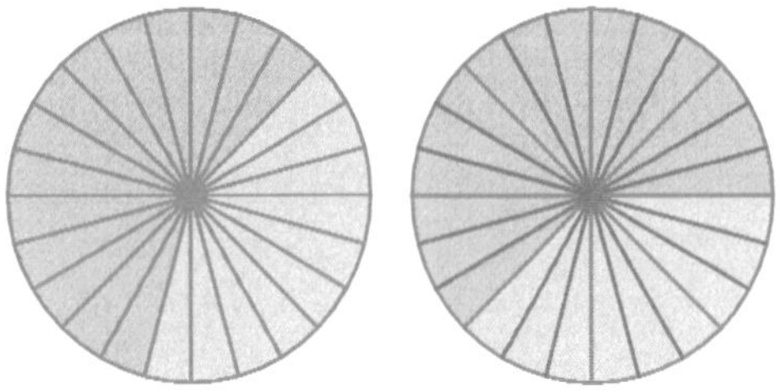 $\frac{14}{24}+\frac{9}{24}=\frac{23}{24}$
36	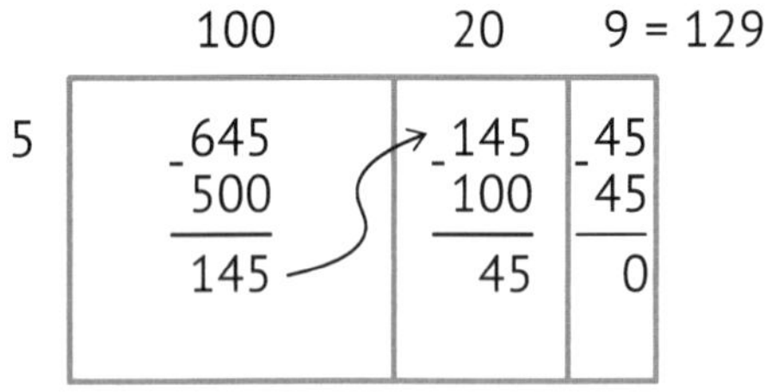
37	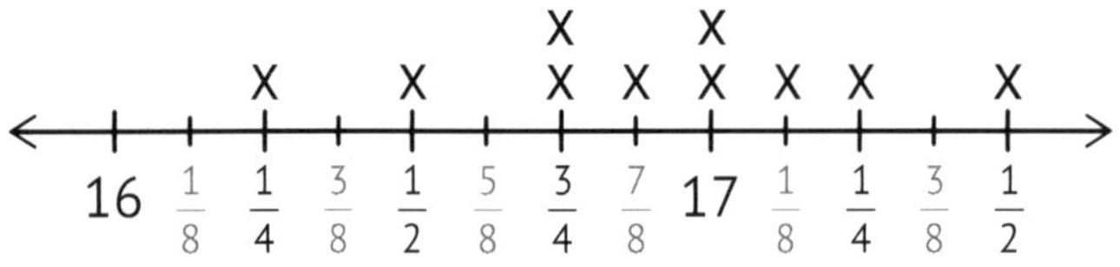
39	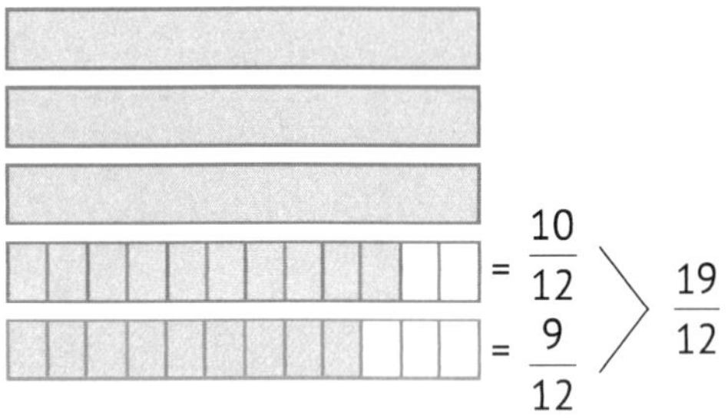 $4\frac{7}{12}$
41	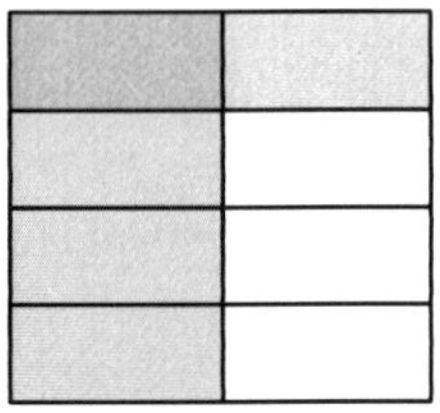

Question	Type	Key	Points	Standard	Cluster	Explanation
Session 1						
1	MC	C	1	CCSS.MATH. CONTENT. 5.NBT.B.5	Numbers and Operations in Base Ten	Students should solve using the traditional algorithm.
2	MC	D	1	CCSS.MATH. CONTENT. 5.NF.B.4	Numbers and Operations Fractions	Count the number of shaded regions to get the numerator 10. Change $\frac{10}{6}$ to a mixed number $1\frac{4}{6}$ candy bars.
3	MC	D	1	CCSS.MATH. CONTENT. 5.MD.A.1	Measurement and Data	Multiply the number of minutes (8) by the number of seconds in one minute (60) to get 480. Add 29 seconds to get 509 seconds.
4	MC	C	1	CCSS.MATH. CONTENT. 5.NBT.A.2	Numbers and Operations in Base Ten	Count the number of zeroes to determine the number of times a decimal must move. When dividing move the decimal to the left. Since the divisor is 1,000, move the decimal to the left three times.
5	MC	D	1	CCSS.MATH. CONTENT. 5.NF.A.2	Numbers and Operations Fractions	Change the denominator to 10. Draw 9 lines to create 10 pieces in the whole. Shade in 8, then subtract 4 to get $\frac{4}{10}$.
6	MC	B	1	CCSS.MATH. CONTENT. 5.MD.C.3	Measurement and Data	The formula for finding the volume of a solid figure is length x width x height.
7	MC	C	1	CCSS.MATH. CONTENT. 5.NBT.B.6	Numbers and Operations in Base Ten	When you subtract you would get 168. Transfer that to the next column. 1 can't be divided by 7, so think of it as 16. You can make 2 groups of 7 with 16, so the next partial quotient is 20
8	MC	B	1	CCSS.MATH. CONTENT. 5.NBT.A.2	Numbers and Operations in Base Ten	Since the product has two zeroes, the second factor must be 100. 10^2 is 100.

GO ON ▶

Question	Type	Key	Points	Standard	Cluster	Explanation
9	MC	C	1	CCSS.MATH. CONTENT. 5.NF.B.6	Numbers and Operations Fractions	Change the mixed number to an improper fraction $\left(\frac{23}{4}\right)$. Multiply that times 6 to determine the area of the picnic area to get $\frac{138}{4}$. Reduce that to $34\frac{2}{4}$ which represents the area. Divide $34\frac{2}{4}$ by 4 to determine the number of blankets she would need. There would be $\frac{2}{4}$ of a foot leftover if she used 8 blankets, so she needs 9.
10	MC	D	1	CCSS.MATH. CONTENT. 5.G.B.3	Geometry	All squares have equal sides (this is what differentiates them from rectangles) All squares are parallelograms because they have parallel sides. All squares have four right angles.
11	MC	A	1	CCSS.MATH. CONTENT. 5.NBT.A.4	Numbers and Operations in Base Ten	The student must recognize that the question is asking about the tenths place rather than the tens place. The digit 1 is in the tenths place. The 4 in the hundredths place indicates that the value is closer to 1 tenth than it is to 2 tenths.
12	MC	C	1	CCSS.MATH. CONTENT. 5.NF.A.2	Numbers and Operations Fractions	You can automatically reason that the sum would be greater than $\frac{1}{2}$ because $\frac{2}{8} + \frac{3}{8}$ would be $\frac{5}{8}$, which is greater than $\frac{1}{2}$. $\frac{5}{8}$ is close to a half, so adding $\frac{1}{4}$ more would not produce more than a whole cup.
13	MC	C	1	CCSS.MATH. CONTENT. 5.MD.C.4	Measurement and Data	Count the length (8), width (2), and height (3). Multiply those dimensions together to determine the volume.
14	MC	A	1	CCSS.MATH. CONTENT. 5.NBT.B.5	Numbers and Operations in Base Ten	Students should solve using the traditional algorithm.

Question	Type	Key	Points	Standard	Cluster	Explanation
15	MC	B	1	CCSS.MATH. CONTENT. 5.NF.B.5	Numbers and Operations Fractions	$\frac{1}{2}$, $\frac{5}{5}$, and $\frac{2}{9}$ are all equal to or less than 1 whole. $\frac{6}{5}$ is greater than 1 whole, so it would yield the greatest product.
16	MC	C	1	CCSS.MATH. CONTENT. 5.NBT.A.3	Numbers and Operations in Base Ten	All of the whole number values are equal, so you have to evaluate the placement of the nine in the decimal. In option C, the nine is in the thousandths place.
17	MC	D	1	CCSS.MATH. CONTENT. 5.NF.B.3	Numbers and Operations Fractions	There are 7 pies that would be divided, so that would be the numerator. There are 8 people wanting to divide the pies, so that would be the denominator.
18	MC	D	1	CCSS.MATH. CONTENT. 5.MD.C.5	Measurement and Data	Multiply the length, width and height to determine the volume. (The information that she needs a box that is 30 cubic inches is extraneous.)
19	MC	B	1	CCSS.MATH. CONTENT. 5.NF.B.7	Numbers and Operations Fractions	The yellow column represents $\frac{1}{5}$. It is divided horizontally into 4 pieces - the top fourth is shaded blue. The overlapping green section represents the quotient. It is one out of 20 pieces.
20	MC	C	1	CCSS.MATH. CONTENT. 5.NBT.A.3	Numbers and Operations in Base Ten	Snowflake C has the smallest width because its value is in the hundredths place, which is less than the tenths place. When compared with snowflake D, .02 is less than .05.
21	MC	A	1	CCSS.MATH. CONTENT. 5.NF.A.1	Numbers and Operations Fractions	Change the denominator to 15 by multiplying the first two fractions by 3 and 5 respectively. Add $\frac{12}{15} + \frac{10}{15} + \frac{7}{15}$ to get the sum $\frac{29}{15}$. Change to a mixed number to get $1\frac{14}{15}$.
22	MC	B	1	CCSS.MATH. CONTENT. 5.MD.A.1	Measurement and Data	Multiply the number of cups (5) by the number of tablespoons in a cup (16). Divide that by 5 to determine that he can fill 16 pouches.

GO ON

Question	Type	Key	Points	Standard	Cluster	Explanation
23	MC	D	1	CCSS.MATH. CONTENT. 5.MD.B.2	Measurement and Data	The longest car is $19\frac{1}{2}$ feet long. The shortest car is 18 feet long. The difference between them is $1\frac{1}{2}$ feet.
24	MC	C	1	CCSS.MATH. CONTENT. 5.MD.B.2	Measurement and Data	He measured 15 cars because there are 15 Xs on the line plot.
25	MC	D	1	CCSS.MATH. CONTENT. 5.NBT.A.1	Numbers and Operations in Base Ten	When dividing by 100 move the decimal point to the left two times.
26	MC	C	1	CCSS.MATH. CONTENT. 5.NBT.B.6	Numbers and Operations in Base Ten	There are 954 tiles to be divided. The product closest to that number is 900.
27	MC	C	1	CCSS.MATH. CONTENT. 5.NF.B.5	Numbers and Operations Fractions	The fractions less than one whole include $\frac{3}{4}$ and $\frac{1}{5}$. $\frac{1}{5}$ is less than $\frac{3}{4}$, so they would go in that order. $\frac{3}{3}$ is equal to one whole, so it would go next. $\frac{9}{5}$ is greater than 1 whole, so it would be last.
28	MC	C	1	CCSS.MATH. CONTENT. 5.NF.B.6	Numbers and Operations Fractions	Multiply the numerator by the whole number to get 72. Divide by 4 to get 18.
29	MC	B	1	CCSS.MATH. CONTENT. 5.OA.A.2	Operations and Algebraic Thinking	The addition segment of the phrase needs to be in parentheses because it does not say "add" it says "the sum of," indicating the addition needs to happen first.
30	MC	B	1	CCSS.MATH. CONTENT. 5.G.B.4	Geometry	The lines indicate that two of the sides are the same length. The curves indicate that the angles are not right angles. They are less than 90 degrees, so they are acute.
Session 2						
31	MC	A	1	CCSS.MATH. CONTENT. 5.NF.B.6	Numbers and Operations Fractions	Multiply the numerators and then the denominators.
32	SR	451 miles each day	2	CCSS.MATH. CONTENT. 5.NBT.B.6	Numbers and Operations in Base Ten	Knowing the math fact 5 x 400 = 2,000 would help him figure out how many groups he could make as he starts to divide. Multiplication is groups repeatedly added to make a product. Division, the inverse, is groups being subtracted to determine the number of groups (quotient).

Question	Type	Key	Points	Standard	Cluster	Explanation
33	MC	D	1	CCSS.MATH. CONTENT. 5.MD.A.1	Measurement and Data	Multiply 14 by 16 to determine the number of oz in 14 lbs. Multiply that product by .02 to determine the amount needed to ship the package.
34	SR	720	2	CCSS.MATH. CONTENT. 5.NBT.B.5	Numbers and Operations in Base Ten	Ramsey accidentally multiplied by 3 instead of 30. He made an error in place value when multiplying.
35	MC	C	1	CCSS.MATH. CONTENT. 5.NBT.B.6	Numbers and Operations in Base Ten	Cross out all seven cubes, but you still need two more. "Break apart" one of the rods by crossing it off, then drawing cubes. Take away 2 cubes. Cross out two rods, but you need to take two more. "Break apart" one of the flats to get ten more rods. Cross off the remaining two rods. Count the number of flats, rods, and cubes that remain.
36	MC	B	1	CCSS.MATH. CONTENT. 5.NF.B.7	Numbers and Operations Fractions	Eight whole figures divided into 4 pieces would yield 32 pieces.
37	SR	$4\frac{1}{8}$	2	CCSS.MATH. CONTENT. 5.NF.A.2	Numbers and Operations Fractions	See figure. Draw a number line with at least 5 whole numbers. Divide each whole number into 4 pieces, creating fourths. Shade $3\frac{3}{4}$ to show the first skip. Divide each fourth into two pieces to create eighths. Shade $\frac{3}{8}$ to show the second skip.
38	SR	Responses may vary. 2 x 3 x 4 1 x 6 x 4 1 x 8 x 3 1 x 12 x 2	1	CCSS.MATH. CONTENT. 5.MD.C.4	Measurement and Data	Three dimensions must be multiplied to create the product 24.
39	ER	2,4,4	2	CCSS.MATH. CONTENT. 5.OA.B.3	Operations and Algebraic Thinking	The rule for determining the x coordinates is to add 1 every time. The rule for determining the y coordinates is to subtract 2 every time.
40	MC	B	1	CCSS.MATH. CONTENT. 5.NBT.B.6	Numbers and Operations in Base Ten	The missing partial quotient is 80 because 6 times 80 is 480.

GO ON

Question	Type	Key	Points	Standard	Cluster	Explanation
41	MC	D	1	CCSS.MATH. CONTENT. 5.NF.A.2	Numbers and Operations Fractions	Change the denominators of the fractions to 12 to get $2\frac{6}{12} + \frac{7}{12} + \frac{8}{12}$ Add to get $2\frac{21}{12}$. Change the mixed fraction to an improper fraction to get $\frac{45}{12}$. This is equivalent to 3.75. Multiply that by two to represent that he has to travel those distances twice in order to make it back home. The answer is 7.5
42	MC	B	1	CCSS.MATH. CONTENT. 5.MD.A.1	Measurement and Data	There are 3 feet in one yard, so divide the number of feet (48) by 3.
43	SR	A. > B. = C. > D. <	2	CCSS.MATH. CONTENT. 5.NBT.A.3	Numbers and Operations in Base Ten	82.15 is greater than 82.105 because the 5 in the hundredths place is greater than 0. 91.1 is equal to 91.10 because the zero in the hundredths place does not change the value of the number. 45.1 is greater than 4.51 because 45 is greater than 4. 92.151 is less than 92.251 because the 1 in the tenths place is less than 2.
44	SR	Oscar	2	CCSS.MATH. CONTENT. 5.NF.B.5	Numbers and Operations Fractions	Possible answer: Oscar is right. Multiplication does not always yield a larger project. When multiplying 723 by $\frac{1}{2}$, you'd essentially be multiplying it by the numerator, then dividing it by the denominator. This would divide 723 into two groups, which would yield a smaller product.
45	MC	B	1	CCSS.MATH. CONTENT. 5.NF.B.6	Numbers and Operations Fractions	Change the mixed number to an improper fraction, the multiply. Change the improper fraction to a mixed number.

Question	Figures
5	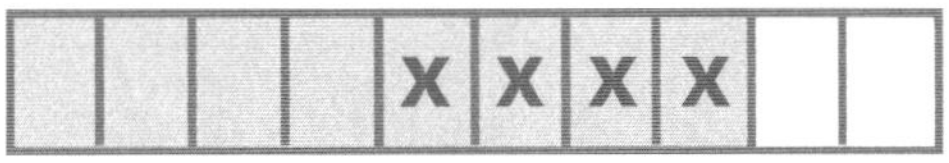 $\frac{8}{10} - \frac{2}{5}$ $\frac{8}{10} - \frac{4}{10}$
37	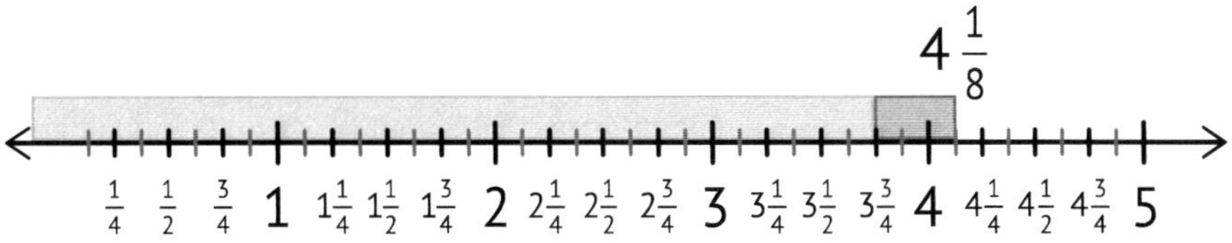

GO ON

Question	Type	Key	Points	Standard	Cluster	Explanation
Session 1						
1	MC	B	1	CCSS.MATH. CONTENT. 5.NF.A.2	Numbers and Operations Fractions	Since $\frac{2}{5}$ is close to a half, adding $\frac{1}{8}$ gets him close to a half. To solve you can change the denominator to 40 to get $\frac{5}{40} + \frac{16}{40}$. This would yield $\frac{21}{40}$, which is almost exactly half.
2	MC	B	1	CCSS.MATH. CONTENT. 5.MD.A.1	Measurement and Data	Multiply 36 x 2 to get 72. That is the total number of ounces she is trying to sell. Divide by 16 to determine the number of lbs. $\frac{72}{16}$ is 4 r 8. Since 8 oz is $\frac{1}{2}$ a lb, the answer is $4\frac{1}{2}$ lbs.
3	MC	A	1	CCSS.MATH. CONTENT. 5.NBT.A.3	Numbers and Operations in Base Ten	When comparing place values, start at the greatest place value. If the values match, continue moving left until one is greater than the other.
4	MC	C	1	CCSS.MATH. CONTENT. 5.NF.B.3	Numbers and Operations Fractions	Each banana would be cut into 6 pieces. Each monkey would get $\frac{1}{5}$ of each banana, so they would have $\frac{5}{6}$.
5	MC	C	1	CCSS.MATH. CONTENT. 5.OA.A.1	Operations and Algebraic Thinking	Multiply first to get 24. Divide $\frac{24}{6}$ to get 4. Add 45 + 4 to get 49.
6	MC	B	1	CCSS.MATH. CONTENT. 5.NBT.B.7	Numbers and Operations in Base Ten	The flats represent the dollars, rods represent dimes, and cubes represent pennies. Draw $3.25 in each box then count the total.
7	MC	A	1	CCSS.MATH. CONTENT. 5.NF.B.7	Numbers and Operations Fractions	The yellow column represents $\frac{1}{6}$. It has been divided into 4 pieces horizontally, so the matching expression is $\frac{1}{6}$ divided by 4.
8	MC	C	1	CCSS.MATH. CONTENT. 5.NF.B.7	Numbers and Operations Fractions	The quotient is related to the size of one piece, so the quotient is 1 out of 24 pieces.

Question	Type	Key	Points	Standard	Cluster	Explanation
9	MC	C	1	CCSS.MATH. CONTENT. 5.NF.A.2	Numbers and Operations Fractions	$\frac{6}{7}$ is close to a whole. $\frac{3}{5}$ is a little more than half. When they are combined, they create a little more than whole. Add $\frac{1}{2}$ to get two books.
10	MC	C	1	CCSS.MATH. CONTENT. 5.NBT.B.6	Numbers and Operations in Base Ten	Since the dividend is closest to 800, the equation that will help her divide would be 4 x 200 = 800.
11	MC	B	1	CCSS.MATH. CONTENT. 5.NBT.B.6	Numbers and Operations in Base Ten	800 divided by 4 is 200. When you subtract to determine how much is yet to be divided, you would get 42. 42 divided by 4 is 10. There would be 2 leftover photos.
12	MC	C	1	CCSS.MATH. CONTENT. 5.MD.C.5	Measurement and Data	Multiply the length (12), depth (3) and height (18) to determine the volume.
13	MC	C	1	CCSS.MATH. CONTENT. 5.NBT.B.7	Numbers and Operations in Base Ten	The flats represent the dollars, rods represent dimes, and cubes represent pennies.
14	MC	B		CCSS.MATH. CONTENT. 5.NBT.B.7	Numbers and Operations in Base Ten	Students can use the visual model or algorithm to add.
15	MC	A	1	CCSS.MATH. CONTENT. 5.MD.C.4	Measurement and Data	Count the cubes to determine the length (10), width (3), and height (5), then multiply the three dimensions to determine the volume.
16	MC	A	1	CCSS.MATH. CONTENT. 5.G.B.3	Geometry	Although some triangles have one obtuse or one right angle, every triangle has at least one acute angle.
17	MC	C	1	CCSS.MATH. CONTENT. 5.MD.B.2	Measurement and Data	There are 18 Xs on the chart representing 18 different trees.
18	MC	D	1	CCSS.MATH. CONTENT. 5.MD.B.2	Measurement and Data	The largest tree is 6 inch wide. The smallest tree is $3\frac{1}{4}$ inches wide. Subtract 6 - $3\frac{1}{4}$ by 'breaking apart' the whole number to become $5\frac{4}{4}$.
19	MC	B	1	CCSS.MATH. CONTENT. 5.NBT.A.2	Numbers and Operations in Base Ten	There are three zeros in the product, so factor 10^3 would yield the product 73,000.

GO ON ▶

Question	Type	Key	Points	Standard	Cluster	Explanation
20	MC	A	1	CCSS.MATH. CONTENT. 5.MD.C.3	Measurement and Data	Inches, feet, and centimeters can all be used to determine the volume because they can be used to measure length, width, and height. Degrees are a measurement of temperature and cannot be used to measure length, width, and height.
21	MC	C	1	CCSS.MATH. CONTENT. 5.NBT.A.4	Numbers and Operations in Base Ten	The number 46.15 would round to 46.2 because the digit in the hundredths place indicates that the number is closer to 46.2 than it is to 46.1.
22	MC	C	1	CCSS.MATH. CONTENT. 5.OA.A.2	Operations and Algebraic Thinking	The phrase "the product of" indicates that this operation should be completed before there rest of the operations.
23	MC	B	1	CCSS.MATH. CONTENT. 5.NBT.A.1	Numbers and Operations in Base Ten	Each place value is ten times greater than the next. 40 is two place values removed from .4, so it is 100 times greater.
24	MC	B	1	CCSS.MATH. CONTENT. 5.NF.A.1	Numbers and Operations Fractions	Change the denominator to 12 then add $\frac{3}{12} + \frac{5}{12}$ to get $\frac{8}{12}$. The question is asking about how much is left to do, so you have to subtract from the whole number. $\frac{12}{12} - \frac{8}{12} = \frac{4}{12}$, which can be simplified to $\frac{1}{3}$.
25	MC	C	1	CCSS.MATH. CONTENT. 5.MD.A.1	Measurement and Data	There are three feet per yard, so there would be 15 groups of 3. Multiply 3 x 15 to get 45.
26	MC	C	1	CCSS.MATH. CONTENT. 5.NBT.B.5	Numbers and Operations in Base Ten	Students should solve using the traditional algorithm.
27	MC	C	1	CCSS.MATH. CONTENT. 5.NF.B.4	Numbers and Operations Fractions	Multiply the whole number by the numerator to get 24. Divide 24 by 5 to get 4 with a remainder of 4.
28	MC	D	1	CCSS.MATH. CONTENT. 5.NF.B.6	Numbers and Operations Fractions	Shade two of the three columns to represent $\frac{2}{3}$. Shade four of the five rows to represent $\frac{4}{5}$. The overlapping shaded region is the product.

Question	Type	Key	Points	Standard	Cluster	Explanation
29	MC	D	1	CCSS.MATH. CONTENT. 5.NF.B.5	Numbers and Operations Fractions	Both $\frac{1}{3}$ and $\frac{2}{5}$ are less than 1 whole, so they would generate lesser products than the factor 54. $\frac{9}{9}$ is equal to 1 whole, so the product would be 54. $\frac{3}{2}$ is greater than 1 whole, so the product would be greater than 54.
30	MC	B	1	CCSS.MATH. CONTENT. 5.NF.B.6	Numbers and Operations Fractions	Change the mixed numbers to improper fractions. $\left(\frac{7}{3}, \frac{29}{5}\right)$ Multiply the numerators together to get 203. Multiply the denominators to get 15. 203 divided by 15 is 13 with a remainder of 8.
Session 2						
31	MC	C	1	CCSS.MATH. CONTENT. 5.NF.B.3	Numbers and Operations Fractions	Each strip of fabric would be cut into 9 pieces. Each student would get one piece from each strip. The number of pieces (8) would become the numerator. The number of students (9) would be the denominator
32	SR	33	2	CCSS.MATH. CONTENT. 5.NBT.A.4	Numbers and Operations in Base Ten	Write the whole numbers at the ends of the line. Divide the line into tenths by drawing 9 tally marks. Place the point at the 7th tally mark to represent 32.7. It is closer to 33, so it would round to 33.
33	MC	C	1	CCSS.MATH. CONTENT. 5.NF.B.7	Numbers and Operations Fractions	The whole would be divided into 4 pieces, creating a denominator of 20 (since it was already cut into fifths). Each friend would get 1 of the four pieces, which would be $\frac{1}{20}$th of the whole.
34	MC	D	1	CCSS.MATH. CONTENT. 5.NBT.A.1	Numbers and Operations in Base Ten	The millions place is four place values removed from the hundreds place. There would be four zeroes in the factor, which would be 10,000.
35	SR	520	2	CCSS.MATH. CONTENT. 5.NBT.B.6	Numbers and Operations in Base Ten	See figure.

GO ON ▶

Question	Type	Key	Points	Standard	Cluster	Explanation
36	SR	Correct	1	CCSS.MATH. CONTENT. 5.NF.B.5	Numbers and Operations Fractions	Any time a whole number is multiplied by a fraction less than one the product is less than the whole number fraction
37	MC	C	1	CCSS.MATH. CONTENT. 5.NBT.B.5	Numbers and Operations in Base Ten	Students should use the traditional algorithm to solve.
38	SR	A. = B. > C. > D. <	2	CCSS.MATH. CONTENT. 5.NBT.A.3	Numbers and Operations in Base Ten	When comparing numbers, start at the greatest place value. If the digits match, keep moving to the next smallest place value until you find digits that do not match.
39	SR	$\frac{4}{15}$	2	CCSS.MATH. CONTENT. 5.NF.B.6	Numbers and Operations Fractions	See figure. Four of the five rows are shaded blue to represent $\frac{4}{5}$. One of the three columns is shaded yellow to represent $\frac{1}{3}$. The overlapping green portion is representative of the numerator (4). The total number of pieces represents the denominator.
40	MC	B	1	CCSS.MATH. CONTENT. 5.OA.A.1	Operations and Algebraic Thinking	Complete the operations in parenthesis first. Then multiply 7 x 9 to get 63. Divide by 3 to get 21.
41	ER	324	2	CCSS.MATH. CONTENT. 5.NBT.B.5	Numbers and Operations in Base Ten	Responses will vary.
42	MC	B	1	CCSS.MATH. CONTENT. 5.MD.A.1	Measurement and Data	Multiply 5 by 16 to determine the number of ounces the shelf could support. Divide by 4 to determine how many books could be supported by the shelf.
43	MC	A	1	CCSS.MATH. CONTENT. 5.NF.A.2	Numbers and Operations Fractions	Change the denominator to 6 to get $\frac{1}{6} + \frac{3}{6}$. Add to get $\frac{4}{6}$, which can be simplified to $\frac{2}{3}$. This represents how much he's already practiced. Subtract from 1 whole $\left(\frac{3}{3}\right)$ to determine how much time he still needs to practice.
44	MC	B	1	CCSS.MATH. CONTENT. 5.NBT.B.6	Numbers and Operations in Base Ten	Solve by using the partial quotient method.

Question	Type	Key	Points	Standard	Cluster	Explanation
45	MC	C	1	CCSS.MATH.CONTENT.5.OA.A.2	Operations and Algebraic Thinking	The parentheses are implied when the phrase "the difference of" indicates that subtraction should have already occurred before the other operations take place.

Question	Figures
32	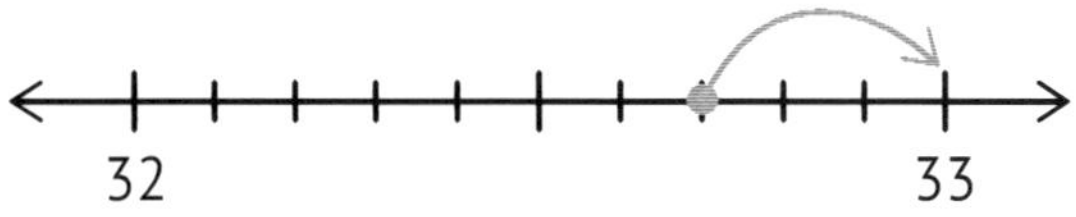
35	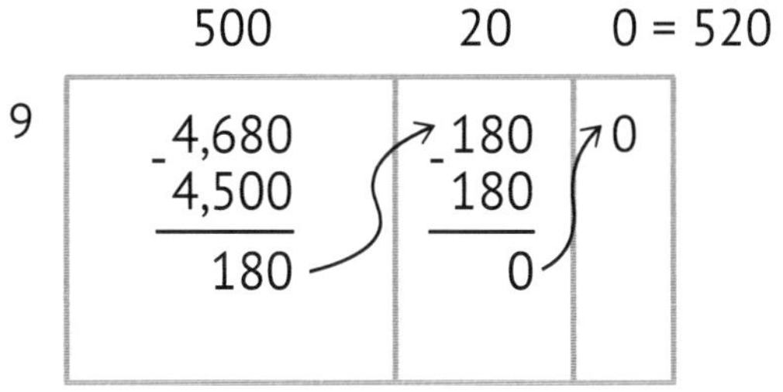
39	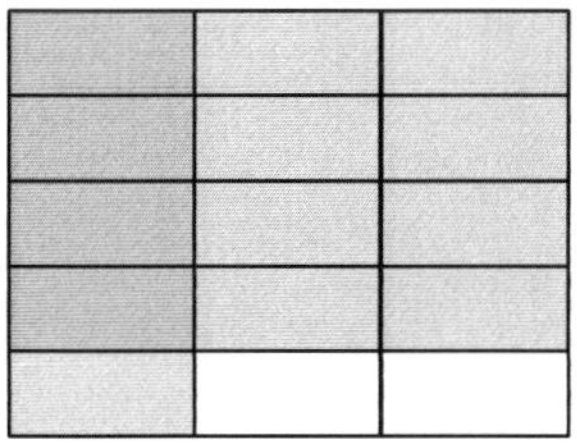

GO ON

Question	Type	Key	Points	Standard	Cluster	Explanation
Session 1						
1	MC	C	1	CCSS.MATH. CONTENT. 5.NBT.B.7	Numbers and Operations in Base Ten	Use the model to subtract by crossing off the blocks starting with the smallest place value. Cross of 5 cubes, 6 rods, and 2 flats. Count the remaining blocks. Flats represent whole numbers, rods represent the tenths place, cubes represent the hundredths place.
2	MC	A	1	CCSS.MATH. CONTENT. 5.NF.B.5	Numbers and Operations Fractions	The greatest product would be 561 x $\frac{7}{5}$ because $\frac{7}{5}$ is greater than 1 whole. Next would be $\frac{8}{8}$ because it is equal to 1 whole. Next would be $\frac{2}{5}$ because it is less than one whole, but it is greater than $\frac{1}{4}$.
3	MC	B	1	CCSS.MATH. CONTENT. 5.NBT.A.4	Numbers and Operations in Base Ten	Because there are 10 tally marks between 23 and 24, they represent the tenths place. The point is at 23.4, which is closer to 23 than it is to 24.
4	MC	B	1	CCSS.MATH. CONTENT. 5.NBT.B.6	Numbers and Operations in Base Ten	When dividing, use a quotient that is less than your dividend, so that you can determine how much is yet to divide. 480 is less than 514, so it is closest to the dividend without going over.
5	MC	D	1	CCSS.MATH. CONTENT. 5.NBT.B.6	Numbers and Operations in Base Ten	514 divided by 12 is 42 with a remainder of 10. Since those samples would need to shipped two, the archeologist would need 43 boxes.
6	MC	B	1	CCSS.MATH. CONTENT. 5.NF.B.3	Numbers and Operations Fractions	The paper would be split into 7 pieces or sevenths. Each person would get $\frac{1}{7}$ from each page, which would add up to $\frac{5}{7}$.
7	MC	C	1	CCSS.MATH. CONTENT. 5.MD.B.2	Measurement and Data	Most of the data points are clustered around 210, making it the average lifespan.

Question	Type	Key	Points	Standard	Cluster	Explanation
8	MC	D	1	CCSS.MATH. CONTENT. 5.MD.B.2	Measurement and Data	Each whole number is divided into 6 parts, meaning each tally mark is representative of $\frac{1}{6}$. The data point that is furthest to the right is above $212\frac{1}{6}$ hrs.
9	MC	A	1	CCSS.MATH. CONTENT. 5.NBT.B.5	Numbers and Operations in Base Ten	Students should solve using the traditional algorithm.
10	MC	B	1	CCSS.MATH. CONTENT. 5.NF.A.2	Numbers and Operations Fractions	$\frac{1}{6}$ and $\frac{1}{8}$ are both smaller than a quarter. When you add them to $\frac{1}{4}$, the sum is slightly larger than $\frac{1}{2}$.
11	MC	C	1	CCSS.MATH. CONTENT. 5.NF.B.6	Numbers and Operations Fractions	The yellow rows represent $\frac{3}{6}$. The blue columns represent $\frac{3}{4}$.
12	MC	A	1	CCSS.MATH. CONTENT. 5.NF.B.6	Numbers and Operations Fractions	The region that is double shaded (green) represents the product.
13	MC	D	1	CCSS.MATH. CONTENT. 5.G.B.3	Geometry	A trapezoid is a type of quadrilateral that has one set of parallel sides. It has two obtuse and two acute angles. Trapezoids do not have equal sides.
14	MC	B	1	CCSS.MATH. CONTENT. 5.NBT.A.3	Numbers and Operations in Base Ten	When reading a decimal, the place value that is identified is where the last digit 'sits.' In this case, in the number 383.083, the 3 sits in the thousandths place.
15	MC	B	1	CCSS.MATH. CONTENT. 5.NF.B.4	Numbers and Operations Fractions	Multiply the numerators to get 3. Multiply the denominators to get 8.
16	MC	B	1	CCSS.MATH. CONTENT. 5.G.B.4	Geometry	It is not a quadrilateral because it does not have 4 sides.
17	MC	D	1	CCSS.MATH. CONTENT. 5.NBT.B.7	Numbers and Operations in Base Ten	Multiply 4.25 x 7 to get 29.75. Subtract 30 - 29.75 to get .25 inches.
18	MC	C	1	CCSS.MATH. CONTENT. 5.NF.A.2	Numbers and Operations Fractions	Change the denominator to 8 by drawing 2 more lines through the circle (dividing each section in half). Shade 5 additional pieces to get $\frac{7}{8}$.

GO ON

Question	Type	Key	Points	Standard	Cluster	Explanation
19	MC	C	1	CCSS.MATH. CONTENT. 5.MD.C.4	Measurement and Data	Count all of the cubes or determine the length, width, and height and then multiply.
20	MC	D	1	CCSS.MATH. CONTENT. 5.NBT.B.6	Numbers and Operations in Base Ten	Students may use an area model or partial quotient method to solve.
21	MC	D	1	CCSS.MATH. CONTENT. 5.NF.B.7	Numbers and Operations Fractions	The 12 pies are divided into 8 pieces each. There are 12 groups of 8 or 96 pieces.
22	MC	C	1	CCSS.MATH. CONTENT. 5.NBT.A.2	Numbers and Operations in Base Ten	Each exponent moves the decimal to the right. Both expressions would yield 42,500, so they are equal.
23	MC	A	1	CCSS.MATH. CONTENT. 5.NF.A.1	Numbers and Operations Fractions	Change the denominator to 24. This would mean the fractions become $\frac{6}{24}$, $\frac{8}{24}$, and $\frac{3}{24}$. Add them together to get $\frac{17}{24}$.
24	MC	C	1	CCSS.MATH. CONTENT. 5.MD.C.5	Measurement and Data	Multiply the length, width, and height to determine the volume of one box - 4 cubic feet. If she has 4 boxes, then she has 16 cubic feet of space.
25	MC	C	1	CCSS.MATH. CONTENT. 5.NBT.A.1	Numbers and Operations in Base Ten	Each place value is ten times greater than the next. 700,000 is three place values removed from 700, so it is 1,000 times greater.
26	MC	D	1	CCSS.MATH. CONTENT. 5.NF.A.2	Numbers and Operations Fractions	Change the denominator to 15 to get $\frac{6}{15}+\frac{5}{15}$. Add to get $\frac{11}{15}$. Since the question was asking about how much is left over, subtract $\frac{11}{15}$ from the whole $\frac{15}{15}$.
27	MC	B	1	CCSS.MATH. CONTENT. 5.NF.B.7	Numbers and Operations Fractions	The $\frac{1}{4}$ piece would be divided into 5 pieces, as would the rest of the whole. This would yield a denominator of 20. The numerator would remain 1 since each family member would get 1 piece out of the whole.

Question	Type	Key	Points	Standard	Cluster	Explanation
28	MC	B	1	CCSS.MATH. CONTENT. 5.MD.A.1	Measurement and Data	Multiply 30 by 2 to get 60. This represents the number of tablespoons he'd consume over 30 days. Divide by 16 to determine the number of cups. $\frac{60}{16}$ is 3 r 12. $\frac{12}{16}$ can be simplified to $\frac{3}{4}$.
29	MC	B	1	CCSS.MATH. CONTENT. 5.MD.C.3	Measurement and Data	The volume is the capacity of the inside of a shape.
30	MC	A	1	CCSS.MATH. CONTENT. 5.MD.A.1	Measurement and Data	Divide 88,000 by 2,000 to determine the number of tons.
Session 2						
31	SR	$\frac{2}{18}$	1	CCSS.MATH. CONTENT. 5.NF.B.4	Numbers and Operations Fractions	See figure
32	MC	B	1	CCSS.MATH. CONTENT. 5.MD.B.2	Measurement and Data	The tallest plant is $3\frac{5}{6}$ feet. The shortest is $2\frac{2}{6}$ feet. Subtract the fractions $\frac{5}{6}-\frac{2}{6}$ to get $\frac{3}{6}$, which simplifies to $\frac{1}{2}$. Subtract the whole numbers 3 - 2 = 1.
33	MC	B	1	CCSS.MATH. CONTENT. 5.NBT.B.5	Numbers and Operations in Base Ten	Students should solve using the traditional algorithm.
34	ER	13	2	CCSS.MATH. CONTENT. 5.NBT.B.6	Numbers and Operations in Base Ten	Responses will vary.
35	MC	C	1	CCSS.MATH. CONTENT. 5.NF.A.2	Numbers and Operations Fractions	$\frac{2}{8}$ is close to a quarter. $\frac{1}{3}$ is between a quarter and a half. Combined they make slightly more than one whole. To solve with computation change the denominators to 24. Add $\frac{8}{24}+\frac{6}{24}$ to get $\frac{14}{24}$, which can simplify to $\frac{7}{12}$.

GO ON ▶

Question	Type	Key	Points	Standard	Cluster	Explanation
36	SR	30 batches	2	CCSS.MATH. CONTENT. 5.NF.B.7	Numbers and Operations Fractions	See figure. Note: Students may also use circular models or a number line.
37	SR	540	2	CCSS.MATH. CONTENT. 5.NBT.B.5	Numbers and Operations in Base Ten	He did not remember to add the new ten he created when he multiplied 5 x 2.
38	MC	A	1	CCSS.MATH. CONTENT. 5.NF.B.6	Numbers and Operations Fractions	Change the mixed number to an improper fraction $\left(\frac{14}{3}\right)$, then multiply to get $\frac{14}{9}$. Change the improper fraction to a mixed number to get $1\frac{5}{9}$.
39	SR	Incorrect	2	CCSS.MATH. CONTENT. 5.MD.C.3	Measurement and Data	Answers may vary. When determining volume, the cubic units should be packed without any gap or overlap. He also appears to have used one level of cubes and may have neglected to measure the height of the box.
40	SR	$3.25	2	CCSS.MATH. CONTENT. 5.NBT.B.7	Numbers and Operations in Base Ten	Students should draw flats to represent whole dollars, rods to represent dimes, and cubes to represent pennies.
41	SR	12	2	CCSS.MATH. CONTENT. 5.NBT.B.6	Numbers and Operations in Base Ten	See figure.
42	MC	D	1	CCSS.MATH. CONTENT. 5.OA.B.3	Operations and Algebraic Thinking	Each time the value of the x axis increases by one, another group of five is added to the y axis.
43	SR	51,426	2	CCSS.MATH. CONTENT. 5.NBT.A.2	Numbers and Operations in Base Ten	When multiplying by factors of 10, count the number of zeros. If you are multiplying and have a decimal, move the decimal to the right once for every zero in your factor of ten. In this case the decimal would move to the right two times.
44	SR	54.1	2	CCSS.MATH. CONTENT. 5.NBT.A.4	Numbers and Operations in Base Ten	The number 54.13 is between 54.1 and 54.2. Write these at the ends of the number line. Draw nine tally marks to divide the number line into ten pieces. These pieces each represent .01. Place the point at the third tally mark, which would represent 54.13. It is closer to 54.1 than it is 54.2, so it would round down.

Question	Type	Key	Points	Standard	Cluster	Explanation
45	MC	B	1	CCSS.MATH. CONTENT. 5.MD.A.1	Measurement and Data	One yard is equal to 3 feet. Divide 18 by 3 to determine the number of yards.

Question	Figures
31	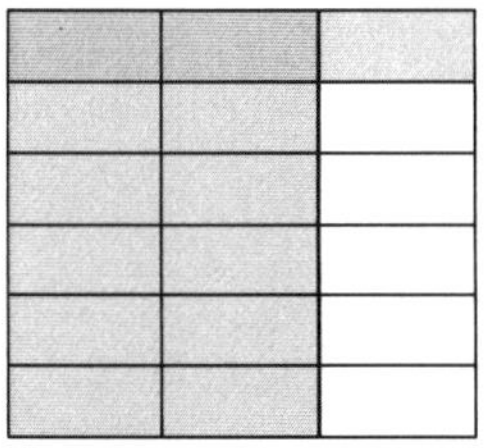
36	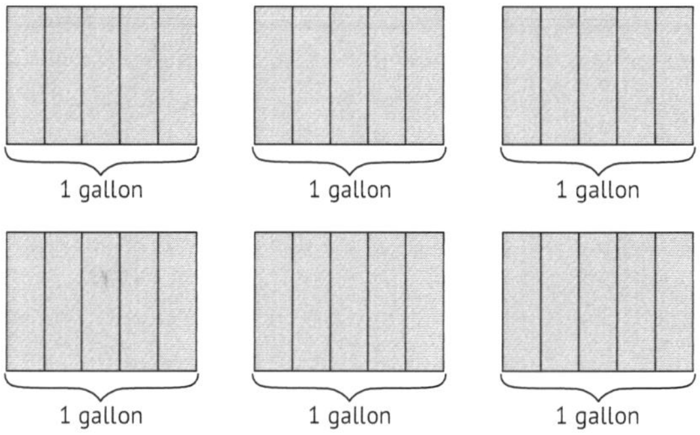
40	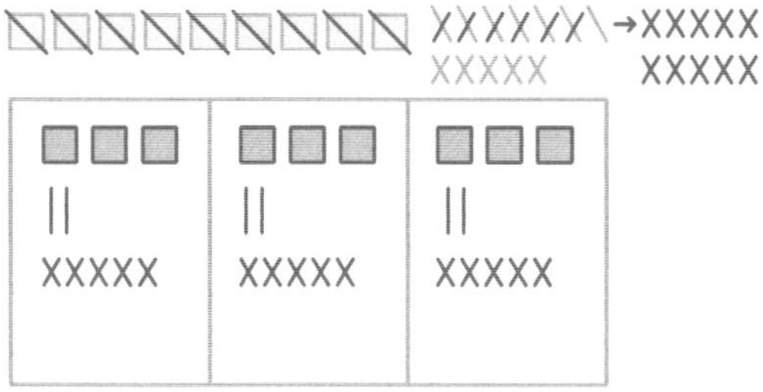
41	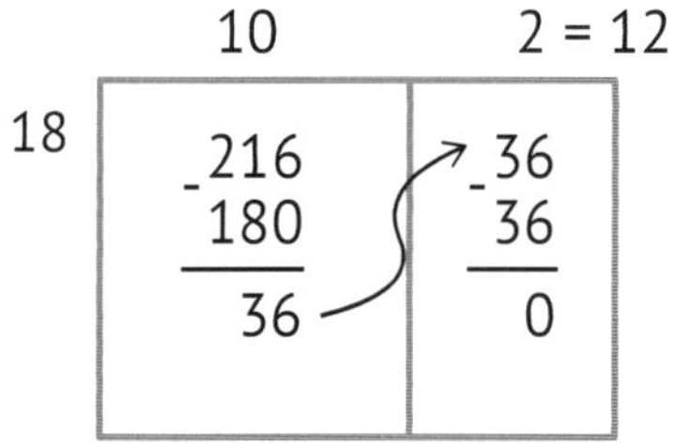
44	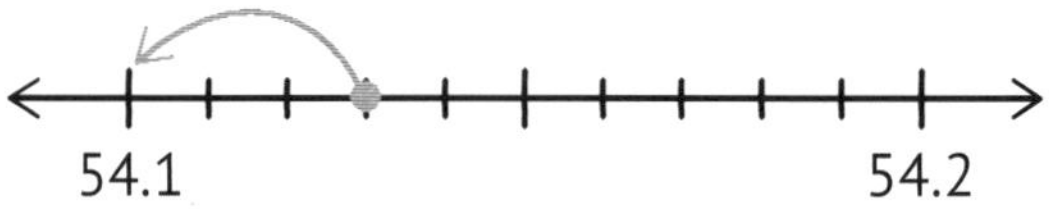

GO ON